GOD AND STEPHEN HAWKING

GOD AND STEPHEN HAWKING

by Robin Hawdon

JOSEF WEINBERGER PLAYS

LONDON

God and Stephen Hawking
First published in 2000
by Josef Weinberger Ltd
12-14 Mortimer Street, London, W1T 3JJ

Copyright © 2000 by Robin Hawdon

The author asserts his moral right to be identified as the author of the work.

ISBN 0 85676 242 3

This play is protected by Copyright. According to Copyright Law, no public performance or reading of a protected play or part of that play may be given without prior authorization from Josef Weinberger Plays, as agent for the Copyright Owners.

From time to time it is necessary to restrict or even withdraw the rights of certain plays. **It is therefore essential to check with us before making a commitment to produce a play.**

NO PERFORMANCE MAY BE GIVEN WITHOUT A LICENCE

AMATEUR PRODUCTIONS
Royalties are due at least fourteen days prior to the first performance. A royalty quotation will be issued upon receipt of the following details:

Name of Licensee
Play Title
Place of Performance
Dates and Number of Performances
Audience Capacity
Ticket Prices

PROFESSIONAL PRODUCTIONS
All enquiries regarding professional rights should be addressed to Alan Brodie Representation, 211 Piccadilly, London W1V 9LD.

OVERSEAS PRODUCTIONS
Applications for productions overseas should be addressed to our local authorised agents. Further details are listed in our catalogue of plays, published every two years, or available from Josef Weinberger Plays at the address above.

CONDITIONS OF SALE
This book is sold subject to the condition that it shall not by way of trade or otherwise be resold, hired out, circulated or distributed without prior consent of the Publisher. **Reproduction of the text either in whole or part and by any means is strictly forbidden.**

Printed by Commercial Colour Press Plc, London E7

GOD AND STEPHEN HAWKING was first presented by Theatre Royal Bath Productions Ltd in association with Derby Playhouse, at the Theatre Royal, Bath on 23rd August 2000, with the following cast:

GOD	Robert Hardy
STEPHEN	Stephen Boxer
JANE HAWKING	Theresa Gallagher
NURSE	Kate Abraham

Directed by Jonathan Church
Designed by Ruari Murchison
Lighting design by Mark Jonathan
Music composed by Matthew Scott

ACKNOWLEDGEMENTS

A large amount of research was necessary for this play. I am indebted in particular to the writings of Kitty Ferguson, Michael Hawkings, John Horgan, J.P. McEvoy and Oscar Zarate, Roger Penrose, Edward Lorenz, William Rankin, Joseph Schwartz and Michael McGuinness, Paul Strathern, John Wheeler, Michael White, John Gribbin, Edward Wilson, John Maddox, Steven Weinberg, Martin Rees, Jim Al-Khalili, Lee Smolin, Ian Stewart and Robert Matthews.

I am grateful for the information provided by Jane Hawking's autobiography *Music To Move The Stars*, and for her comments on the text.

My gratitude also to Professor Mark Birkinshaw of Bristol University, who has modified many of my inexpert interpretations of modern scientific theory.

And of course I am most indebted to the work and writings of Stephen Hawking himself.

This play is a fictional account, and does not necessarily represent the opinions or actions of any person depicted in it.

Robin Hawdon
August 2000

CHARACTERS

GOD   (also seen as NEWTON, EINSTEIN, POPE JOHN PAUL II, THE QUEEN, VARIOUS PROFESSORS and SCIENTISTS, and other assorted characters)

STEPHEN HAWKING

JANE HAWKING

NURSE

SETTING

A bare stage, backed by as large a projection cyclorama as is feasible. This could be mounted above a raised platform or steps to provide a variation to the acting area. A light, movable chair and perhaps a portable lectern are the only furniture.

NB: The projection slides used in the original production of GOD AND STEPHEN HAWKING are available on hire for productions of the play. Please contact Josef Weinberger Ltd for further details.

ACT ONE

The stage is completely bare, but backed by a huge projected panorama of the sky at night. The Milky Way. We gaze at it for several moments.

A figure materialises from the background, clad in a voluminous white robe. He walks to the forestage and addresses the audience.

GOD   Good evening. May I introduce myself. I'm God. (*Beams cheerfully at the audience.*) Interesting that you find that amusing rather than startling or awe-inspiring. Anyway . . . for the purposes of the moment at any rate, I'm God. I'm seen in many guises, but so as not to terrify you too much, I thought I'd start with something fairly mundane. Now one of the aims of tonight's experience is to try and decide whether or not in fact there is a God. If indeed it turns out that we prove there isn't, then I shall disappear in a puff of smoke and you will have to forget I was ever here. (*A beat.*) Pretty illogical you might think. But is there an alternative possibility? That I'm both here and not here at the same time. No, I'm not being frivolous. There is a scientific analogy to this. A well-known proposition known as the case of Schrödinger's cat. Yes, yes, I know it's all too familiar to the students of quantum physics amongst you, but it's very relevant here. So to ring the changes, let's for the hell of it talk about Schrödinger's rabbit instead. It is an extraordinary fact, which bemuses even your own scientists, that an elementary particle – for example a photon of light – can appear to take a number of different paths through space simultaneously, whilst still retaining its properties as a single entity. If therefore in experiment such a particle is emitted which on one of its possible routes could trigger a device which kills a rabbit . . .

(*A diagram of the Schrödinger experiment, using a gun and a rabbit in a cage, appears on the backscreen.*)

. . . and which, on another simultaneous route could harmlessly bypass the device – then that rabbit, in quantum physics terms, is both alive and dead at the same time. In what you might say are parallel versions of the Universe.

(*The rabbit, appears both dead and alive on the diagram. He grins cheerfully.*)

Perplexing – eh? I hasten to add, before the animal rights activists amongst you rise up in arms, that Mr Schrödinger who postulated this very observable paradox, was a humane man and did not actually murder any animals in the course of his experiments. Now – replace the rabbit with me, and you have the situation that I could both exist and not exist in alternative versions of the Universe. All we have to discover is which is the more likely. Or whether, as in the case of the rabbit, *both* are possible. (*A beat.*) Oh dear – perhaps we'd better move on before I confuse myself at the outset.

(*The backscreen reverts to the Milky Way.*)

I want you to meet someone who knows as much as anyone . . . except me – if I exist . . . about the origins of the Universe, the nature of space and time, and all that stuff. He, you might say, is representative of those who certainly *don't* believe in my existence. And for that reason I'm afraid I can't converse with him directly – which gives us a few more problems. (*Winks.*) But I have ways round that.

(*He steps to one side of the stage. There is a whirring sound and a spotlight illuminates an electric wheelchair coming in from the other side at full throttle. It does a smart right turn, rolls to the centre forestage and stops. The familiar distorted figure of* STEPHEN HAWKING *sits hunched within it, surrounded by his computers and buttons. Another moment of stillness. His left fingers move minutely. The famous synthesised voice speaks.*)

STEPHEN   For this final lecture of the academic year I want to pose you a question. Are we close to reaching what some scientists have called the end of science? (*Pause.*) There are two quite distinct reasons for thinking that in pure physics terms we may be. One is that it is conceivable that the final unifying formula which will explain the workings of the Universe – the so-called Theory of Everything – is within our grasp. As we have seen, we have already condensed the various rules which govern the behaviour of matter and energy down to a handful of basic formulae.

(*The backscreen changes to show a mass of complex mathematical calculations.*)

If only our household budgets or end-of-year tax returns could be so simply stated.

(*Grins widely. It is virtually the only other movement he can make.*)

Many scientists, of whom I am one, are optimistic that within the next generation or so we will combine even these principles into one revelatory concept which will explain the Universe to itself. As the famous physicist John Wheeler of Princeton described it: '. . . an idea so simple, so beautiful, so compelling that when we grasp it, we will say to each other, how could it have been otherwise? How could we have been so stupid for so long?'

(*The back screen changes to show several of the great epoch-making formulae – the Pythagoras Triangle; the Pi Equation: $A = pi\ R^2$; Newton's Law of Universal Gravitation: $F_G = -\ Gm,m_2/r^2$;*

Einstein's Theory: E = mc²; Planck's Equation: E = HV; the Vacuum Solution: Rik = 0; the Schwartzchild Radius: R = 2GM/c^2; the Theory of Supergravity: N = 8; the Hawking Formula for Black Hole Radiation: S = KA.)

The other, quite contradictory reason for believing that science might have nowhere much left to progress, is that we have reached the stage where it's no longer possible to ratify many of our conclusions. In the realm of quantum physics we have mathematically traced the essential particles of matter down to such minute and elementary constituents that we cannot actually perceive them by experimental means. And in the realm of cosmology the problem is opposite but similar. We have worked out what we think happens at the very edges of time and space, but it is so far beyond our abilities to actually *observe* that we are left with nothing but a mass of calculations.

(More reams of mathematics behind him.)

Which may give great satisfaction to us boffins but is pretty unsatisfying to the man in the street. However, I want to remind you of a few simple facts. We in our field are so familiar with these things that I think we sometimes forget the wonder of them.

(The backscreen shows a diagram of the solar system.)

Our tiny planet is spinning in its insignificant solar system somewhere on the outer edges of a galaxy that for obvious reasons we call the Milky Way. There are roughly a hundred billion stars in the Milky Way, many with their own attendant planets. Think about that number. One hundred billion. A hundred thousand million. The human brain can barely encompass such a sum. And yet our galaxy is but one of a hundred . . . billion . . . other galaxies.

(Image of the Milky Way. Pauses to let it sink in.)

The light from the nearest star outside our solar system travels four light years to reach us. We all here use the term light year so casually that we forget how far it really is. The distance light travels in a year, moving at three hundred thousand kilometres *per second*. The farthermost stars in the Universe we now know are several billion light years away. At least we know they were there billions of years ago, because that is how we observe them now. What has happened to them in the intervening billions of years we can hypothesise, but we can't actually prove by observation. For all we know they may actually have turned round and be rushing back towards us, they may have exploded in a flash of cosmological fireworks, they may be performing an astral polka together. That, therefore, is the problem. Everything is either too small, or too big, to get a grasp on. As an actress probably once said to a bishop somewhere. Why is it that all the best jokes are sexual? – now there's a worthwhile subject for a thesis.

(Grins.)

So you see what I mean by the end of science. We will either soon know the basis for everything – the final grand theory – or we will always know nothing, because there's no way of proving it. Something of a dilemma! Yet all this has really only happened in the brief spark of time since I entered university forty years ago . . .

(The lights change to brighten the whole stage as Stephen *wheels his chair away to the side, climbs out of it, and stands as a twenty-year-old Oxford undergraduate. He starts to put on a rowing cap and sports shoes. The backscreen shows the architecture of University College, Oxford. At the same time* God, *who has been discreetly listening from the side, rouses himself. He takes off his*

robe, turning it inside out to reveal it as a black don's gown.)

GOD   (*re-gowning himself*) This is where I get to have a bit of fun. We're going into another time dimension, and I'm going into another character dimension. For my own devious purposes, I'm now taking the guise of Stephen Hawking's first physics tutor at Oxford. A man probably not up to his pupil's intellectual capacity, but considerably wiser than him in other ways.

(Takes a sheaf of papers from the folds of the gown.)

I've always seen myself as a player of parts.

(Goes briskly towards STEPHEN *waving the papers.)*

BERMAN   Mr Hawking, Mr Hawking – what is this? I asked you for an essay on the Newtonian Concept of Mass, not a critique of your text books!

STEPHEN   Sorry, Doctor Berman. It was more fun finding the mistakes in them than fiddling about with dear old Newton once again. I mean he does hit the Earth with quite a dull thud himself, doesn't he?

BERMAN   (*sighing*) You're going to come a cropper yourself if you don't take things a *little* more seriously, Mr Hawking. You're a lively chap – I'm sure you keep your friends endlessly amused with all your witty parodies of us lesser mortals. However, it took Newton half a lifetime to work out his theories on gravity, so you're going to have to do a little more than this if you want to get a decent degree.

STEPHEN   Yes.

BERMAN   You never take any notes during our tutorials. Why not?

STEPHEN   Well, er . . . like the textbooks – there doesn't
           seem much point writing things down until one is
           convinced they're right.

BERMAN   Oh, you're challenging *my* teachings now?

STEPHEN   No, no . . . it's just – everything seems so
           incomplete still.

BERMAN   That's the nature of science. If we knew
           everything there'd be nowhere to go.

STEPHEN   Well, that's it. I still don't know where I'm going.

BERMAN   Nor did Newton – until he let his observations tell
           him.

           (*Hands back the papers.* STEPHEN *drops them
           clumsily.*)

           There, you see. . .

STEPHEN   (*frowning*) Sorry. I can't seem to . . . (*Stoops to
           pick them up.*)

BERMAN   You're subject to his laws just like the rest of us.
           You're a clever lad, Stephen, but it won't all come
           by divine inspiration.

STEPHEN   No. I don't believe in divinity anyway.

BERMAN   Ah. Now if you were reading theology, then we
           really could have an interesting debate. However,
           you've settled for more terrestrial matters, so for
           now you must accept that I am your God, and
           though I may work in mysterious ways they are
           intended to lead you into the paths of
           righteousness.

STEPHEN   Yes. I'm afraid I must go, Doctor Berman. I'm late
           for rowing practice.

BERMAN

Ah, yes – you're coxing the college eight, aren't you? I hear you take a somewhat hazardous line with them too sometimes.

(STEPHEN *grins and goes. The backscreen reverts to the Milky Way.* GOD *turns to the audience.*)

GOD

You see what I'm up against? He is representative of my problem. The march of modern science is phenomenal I have to admit, but it is rather leaving me out on a limb. I need all my resources to fight back.

(JANE *enters. She gazes at the heavens.* GOD *looks at her, then knowingly at the audience.*)

Sex. Now there's a great weapon. Did I contrive such a bizarre procreative method deliberately in order to cause mayhem amongst your species? Or was it simply a grotesque effect of the primeval chemistry? Well, wherever it came from I don't see why I shouldn't try and use it to my advantage now. (*Indicates* JANE.) I can converse with her because she believes in me implicitly.

(*Goes towards her. She turns and smiles as if he is an old acquaintance. He indicates the heavens.*)

Wonderful, isn't it?

JANE

Yes.

GOD

How old are you, my child?

JANE

Eighteen.

GOD

Yes – a splendid age.

JANE

It doesn't feel very splendid. In fact it feels very . . .

GOD

What?

JANE

Frustrating.

(*He glances slyly at the audience.*)

GOD   Perhaps I can help.

JANE   Can you?

GOD   Let me pose you a question. Which would you rather? A conventional life to suit your upbringing – few traumas, steady husband, family life – a peaceful end. Or something far more perilous.

JANE   Perilous?

GOD   (*indicating the sky*) A path to the heavens. Traumatic, exhausting – hugely dangerous. But with the chance of glory.

JANE   That, naturally.

GOD   Not naturally at all. Most people would balk when it came to it.

JANE   Not me.

GOD   Well then, perhaps you're the one. Mind – I won't be able to interfere once you've started.

             (*He flicks his fingers as he leaves, and the backscreen changes to show colourful decorations and party lights. Early sixties pop music plays.* STEPHEN *enters in a black velvet jacket and red velvet bow-tie. He carries a glass of wine.* JANE *and he gyrate together.*)

STEPHEN   Thank you for coming.

JANE   You didn't tell me it was your twenty-first.

STEPHEN   I don't like admitting I'm so old.

JANE   Old!

STEPHEN   I've still no idea what to do with myself.

JANE   I hear you got a first-class honours. Not a bad
             start.

STEPHEN   It's where to go next. Physics has so many
             directions.

             (*The music stops. They move to one side.*)

JANE   It's all way above my head.

STEPHEN   Yes – that's probably the one.

JANE   (*puzzled*) What?

             (*He points upwards.*)

STEPHEN   Direction. I suspect cosmology's the place.

JANE   What exactly *is* cosmology?

STEPHEN   Like astronomy – only more. It's why I've left
             Oxford for Cambridge. They're leading the world
             there. Universe I should say.

JANE   (*smiling*) You're funny.

STEPHEN   What are you going to do?

JANE   I'm going to London – languages.

STEPHEN   Worlds apart.

JANE   (*shyly flirtatious*) Well that shouldn't deter a
             cosmologist.

STEPHEN   (*taken aback*) Oh – well . . . (*Goes to drink but
             fumbles with his glass and spills.*) Damn! What's
             the matter with me?

             (*She takes his handkerchief and wipes the front of
             his jacket.*)

JANE   You're pretty clumsy for a scientist.

STEPHEN   It's not just . . . I've got to go for some tests.

JANE   (*concerned*) Tests?

STEPHEN   Oh, nothing serious. Just my coordination's sometimes off.

JANE   (*daring*) Well you'd better get that right if we're going to see much more of each other.

(*Goes. He stands looking after her. The backscreen changes to show the interior of a hospital. A* NURSE *enters.*)

NURSE   Oh, Mr Hawking. The doctor will see you now.

(*She stands on one side.* GOD *enters wearing a white doctor's coat, carrying a medical file. Smiles at the audience as he comes.*)

GOD   God, the healer. Though not always. (*To* STEPHEN, *as* DOCTOR.) Ah yes; good morning.

STEPHEN   You have my tests, Doctor.

DOCTOR   Yes. (*Studying the file.*) You're twenty-one, Mr Hawking?

STEPHEN   Yes.

DOCTOR   And what were your plans?

STEPHEN   Plans?

DOCTOR   For the future.

STEPHEN   I'm doing a PhD at Cambridge. Beyond that . . . (*Shrugs.*)

DOCTOR   Ah.

STEPHEN   Why?

DOCTOR It's not good news, I'm afraid. (*Pause.*) You have something called amyotrophic lateral sclerosis. Better known as motor neurone disease.

STEPHEN What does that mean?

DOCTOR It's a gradual disintegration of the nerve cells controlling muscle activity.

STEPHEN Which muscles?

DOCTOR All muscles. First the major ones associated with body movement, then those affecting speech, swallowing, ultimately breathing. We haven't yet found a cure. It's quite rare – a gradual . . .

STEPHEN How long? (*Pause.*) How long do I have?

DOCTOR The average survival rate at your stage is two to three years.

(*Pause.*)

I'm sorry. I wish I could give you more hope, but . . .

STEPHEN Thank you, Doctor. (*Turns away.*)

DOCTOR The one consolation is that the brain is quite unaffected.

(STEPHEN *goes to one side and sits. The* DOCTOR *calls after him.*)

The secret is not to give in.

(*Turns to the audience taking off his doctor's coat.*)

GOD Oh yes, I know what you're thinking. I am a cruel God – if I exist. Why on this Earth would I permit such aberrations if I did indeed create it. Well, as has already been noted, if I exist I move in mysterious ways . . .

(*Smooths down his hair. Hands his white coat to the* NURSE *with a smile.*)

Thank you.

(*She goes off. He watches her as she goes.*)

She's an angel.

(*Turns back.*)

I'm moving right now to Trinity Hall College, Cambridge . . .

(*The backscreen shows Trinity Hall.*)

. . . or rather I'm moving you there – I'm everywhere . . . if I'm anywhere. Orbiting there is one of the most influential planets in the sparkling firmament of international science – just the sort of wise mentor that's required at this stage.

(STEPHEN *is drinking and listening to Wagner.* GOD, *as* DENNIS SCIAMA, *goes to him and puts an arm round his shoulders.*)

SCIAMA   Stephen, it's no use hiding away all day long with vodka and Wagner, feeling sorry for yourself. The doctors themselves have said – the best thing you can do is get on with your PhD.

STEPHEN   Doctor Sciama, I don't know what to *do* for my PhD! Everything's being covered, by far better brains than mine.

SCIAMA   As to everything being covered, that's highly unlikely. We're only just breaking through to the outermost boundaries of space and time. We're floundering around there like sprats in the sea. As to you're not having the brains, from what I've seen it's not those you lack – it's focus.

STEPHEN   (*despairing*) What's the point?

SCIAMA   Stephen . . .

STEPHEN   You know I was born on the three-hundredth
            anniversary of Galileo's death?

SCIAMA   Really?

STEPHEN   Three hundred years to get from him to here. What
            can I do in two?

SCIAMA   Time is relative, Stephen. Study your Einstein.
            And your Newton, too . . .

STEPHEN   (*sighing*) Not him again.

SCIAMA   Strangely, he was born that same year that Galileo
            died – isn't that auspicious?

STEPHEN   (*unimpressed*) Oh.

SCIAMA   Well, as Newton postulated, an accelerating body
            will go on accelerating unless there's a change in
            the forces affecting it.

STEPHEN   I'm hardly what you'd call an accelerating body.

SCIAMA   Ah, but an accelerating *mind* – that's the thing.
            Isn't it fortuitous that you've chosen theoretical
            physics. The one subject which does not require
            any physical aptitude on your part.

            (JANE *enters in a ball gown.* STEPHEN *sees her and
            smiles.* SCIAMA *looks round. They get up.*)

            More distractions. How serious is this?

STEPHEN   (*shrugging*) We don't know each other very well.

SCIAMA   Are you bringing her to the May Ball?

STEPHEN   Yes.

SCIAMA   Well, that's a positive move. But remember –
          you're studying theoretical physics, not practical
          biology.

          (*Goes.* JANE *runs to* STEPHEN *and hugs him. Then
          looks into his face, concerned.*)

JANE   Are you all right?

STEPHEN   I'm angry.

JANE   (*holding him*) Oh, Stephen . . .

STEPHEN   Oh, not about dying. Just not having time to live.

JANE   I wish you could pray. God would show you the
          way – if you'd let him.

STEPHEN   (*sighing*) Jane, that's all such nonsense. God
          hasn't interfered in the workings of the Universe
          for ten billion years. Why should he do so for me?

JANE   How do you know he hasn't?

STEPHEN   Well, if he has, he's made a pretty rotten job of it!

JANE   I don't understand . . .

STEPHEN   What?

JANE   How you can know so much about the wonders
          out there, and yet not believe in God.

STEPHEN   Precisely *because* I know so much. We've learned
          most of the basic facts about the Universe. We
          don't *need* a God to explain it all now.

JANE   You need a God to explain how it started. Even
          your great Isaac Newton believed that.

STEPHEN   How do you know that?

JANE   I've been reading up on him.

STEPHEN   (*surprised*) Have you now?

JANE   Flattered? Anyway he was obsessed with
              theology. He even learned Hebrew so he could
              study the Bible from the original texts.

STEPHEN   Because he couldn't reconcile it with his science.

JANE   No!

STEPHEN   You know he rejected Catholicism?

JANE   Yes, but . . .

STEPHEN   He even refused the last sacrament on his death
              bed. That's not a man who truly believes in God.

              (GOD *enters in seventeenth-century costume,
              holding a wig.*)

GOD   (*to the audience*) I really can't allow this to go on!
              (*To* STEPHEN.) I'm sorry, Stephen, but I have to
              challenge you on that.

STEPHEN   Doctor Sciama – what on earth . . . ?

SCIAMA   Oh, this year the dons have decided to go to the
              ball in costume. (*Poses.*) What do you think,
              young lady?

JANE   Very splendid.

SCIAMA   And as I'm meant to *be* Isaac Newton it's quite
              fortuitous.

STEPHEN   Ha!

SCIAMA   Now I heard what you said, Stephen, and Newton
              would have answered you quite simply: the more
              he understood of the Universe the more he
              appreciated how miraculous it is.

STEPHEN   No, no! He was the first to really start *exploding*
              its miraculousness. Him and his tedious old

gravitational force. He was the first to show there's nothing divine about the heavens at all!

SCIAMA   Oh, dear boy, don't be so naive. His discoveries merely laid bare the yet greater marvel that lies behind it all. (*Looks surreptitiously about.*) Tell you what – it'll help me get into character . . . (*Mischievously.*) Pretend I'm him.

STEPHEN   What?

SCIAMA   Go on – I challenge you to expose my disbelief in God.

STEPHEN   You don't know how he really thought – nobody does.

SCIAMA   Oh, I think I have a fair idea.

STEPHEN   Now *you're* acting as God.

SCIAMA   (*with a smile*) Humour me. I am Sir Isaac Newton, President of the Royal Society, physicist, mathematician, astronomer, alchemist, optician, chronologist, inventor, philosopher – the most distinguished practitioner of all the sciences who ever lived . . . (*To* JANE.) If a rather objectionable man.

           (*We see the front of Newton's 'Principia Mathematica'. He puts on his wig.*)

           Challenge me. Go on – challenge me.

STEPHEN   This is ridiculous.

SCIAMA   Precisely what they first said about his gravitational attraction.

STEPHEN   Very well, Sir Isaac . . . (*Thinks.*)

SCIAMA   This is quite a fun idea – I might use it in my tutorials. (*To* STEPHEN, *imperious.*) I'm waiting.

STEPHEN All right . . . for the first time in the history of man you have explained in purely technical, mathematical terms the basic workings of the heavens. You have confirmed what Copernicus and Galileo scarcely dared to postulate – that the Earth is not at the centre of the Universe, but a microscopic speck trapped amidst the cosmic litter by virtue of calculable forces exerted on it by all the other specks.

NEWTON Even so.

STEPHEN And that on this spinning speck an astounding but as yet pretty confused form of life has evolved . . .

NEWTON (*wagging a finger*) Tut, tut – Darwin has not made his entrance yet. You may not use such terms.

STEPHEN All right – has arrived somehow. Now this life form is making a pretty good cock-up of everything. It squabbles and fights, it steals its neighbour's asses and seduces his wives, it makes graven images and practices all sorts of bizarre witch-craft.

NEWTON What is your point, sir?

STEPHEN So the theory has arisen that the invisible being who dreamt up this vast and astounding complexity decides he's got it a bit wrong on this particular grain of the cosmos, and he'd better do some tinkering.

NEWTON (*amused*) Tinkering – hah! I like that.

STEPHEN He therefore concocts a scheme to transform himself into human shape, land himself at a completely random point in history in an obscure region far from the centre of civilisation, preach his message to whatever ragged, superstitious tribe happens to inhabit the spot, pull off a few faith-healing stunts, and then hope that by getting himself bumped off at an early age they'll all go bananas over it and spread the word around the rest of the globe.

JANE   (*protesting*) Oh, Stephen . . .

STEPHEN   Now surely, Sir Isaac, your brilliantly analytical mind isn't going to accept that apocryphal piece of mythology?

NEWTON   Indeed no.

JANE   What?

STEPHEN   There.

NEWTON   Of course I do not! My dear child, you should read what I myself have written. It was not Catholicism alone I rejected.

JANE   No?

STEPHEN   No.

NEWTON   I rejected *all* Christian dogma. As you rightly state, I spent much valuable time exploring what was contradictory within the ancient texts of the Bible.

STEPHEN   Then what are we arguing about for Ch . . . ?

NEWTON   You were about to say 'for Christ's sake' then.

(STEPHEN *looks sheepish.* NEWTON *turns to* JANE.)

It's not surprising the various parts of the New Testament conflict – they were all based on anecdote. And written and rewritten over four centuries.

JANE   Really?

NEWTON   All that resurrection stuff – highly improbable. But the idea of resurrection is a sure path to the creation of a divinity.

JANE   (*dismayed*) Oh . . . but surely . . .

NEWTON   What? Do I challenge your deepest beliefs, young
            lady?

JANE   Well, yes. The resurrection's been the cornerstone
            of . . . of . . . Christian faith throughout history.

NEWTON   Yes, it amazes me yet that people will cling to such
            an unlikely fable. But the evidence for it is hardly
            something a modern court of law would accept.

JANE   Well, I . . .

NEWTON   And certainly not a scientist like myself. You
            know, a body can be spirited away in a dozen more
            credible ways.

STEPHEN   I don't understand what your argument is.

NEWTON   My good fellow! Belief in God is a far more
            fundamental business than paying homage to
            ancient legends and superstitions. Christ may
            have been simply a fine philosopher and a brilliant
            hustings orator, and the Bible may have been a
            wondrously constructed work of art, but they both
            carried a pretty sound message.

STEPHEN   I grant that, but . . .

NEWTON   As did Mohammed and the Buddhas and all well-
            established religions throughout history. Why
            may not all the assorted creeds be examples of
            man's fumbling searches for the ultimate truth – as
            are all his flawed scientific theories through the
            ages?

STEPHEN   Because the two are mutually incompatible.

NEWTON   How so?

STEPHEN   The more one dispels the mystery of the Universe,
            the less one has need of a mystical creator.

NEWTON   Foolish boy – to know how is not to know why!
            The more questions we scientists answer, the more
            spring up to take their place. You see, you must
            ask yourself *why* I was so obsessed with the Bible.
            Why I went to my death refusing the last
            sacrament. If I wasn't still struggling with my
            essential belief in God I would not have given a fig
            for such matters!

STEPHEN   You were still struggling for the very reason that
            you didn't yet have all the answers.

NEWTON   And now you do?

STEPHEN   We have more of them. (*Pointedly*.) As you are
            teaching me, Doctor Sciama.

SCIAMA   (*reverting to character, caught out*) Ah . . . yes,
            well . . . that's another matter.

STEPHEN   But relevant, wouldn't you say? Sorry, sir, I'd love
            to argue all night but we must get ready for the
            ball.

            (*He leads* JANE *off by the hand.* SCIAMA *calls after
            them.*)

SCIAMA   Careful now. Don't get bewitched by the occasion.

            (JANE *waves to him as she goes.* GOD *turns out
            front.*)

GOD   Round one – a draw perhaps. And if we've
            offended anyone with our rejection of certain
            widely held doctrines, you must remember that I
            have to debate with him on his own strictly logical
            terms. (*A beat*.) For the moment.

            (*Takes off his wig and mops his brow*.)

            Whew! (*Stares at the wig*.) The vanity of your
            breed! To endure such discomfort for the sake of
            appearances.

(*Starts to undo his coat buttons.*)

Who imbued you with such a quirk? I'm sure it wasn't me.

(*Goes off. The backscreen changes to show part of the gardens. Moonlight, lawns, trees, coloured lights. Dance music plays – Glen Miller or some such.* STEPHEN *and* JANE *enter hand in hand.*)

JANE   A wonderful night.

STEPHEN   I'd rather be at the opera.

JANE   (*slapping him playfully*) You and your opera. Wagner's *your* god, isn't he?

STEPHEN   Well he's certainly provided more conclusive inspiration than most gods.

JANE   Ah, but who gave him the inspiration?

STEPHEN   Science will discover even that one day.

JANE   You really believe that, don't you?

STEPHEN   Yes.

JANE   Well, I don't know who gave Glen Miller his inspiration, but I want to dance.

STEPHEN   I can't dance.

JANE   Of course you can.

(*She takes his arms and leads him into a gentle dance. They sway together for a moment.*)

Oh, Stephen, we're so different – why do we get on so well?

STEPHEN   Opposites attract. Basic law of electro-magnetism. (*He nuzzles her neck.*) Even though you won't come to bed with me.

JANE   I want to. I just . . . Ask me to marry you.

STEPHEN   Jane – you know I . . .

JANE   I don't care how long you've got. Let's make the most of it.

STEPHEN   I'm getting worse. I'm going to be crippled, for God's sake!

JANE   For God's sake?

STEPHEN   No. (*Bitter.*) God obviously gave up on me, didn't he?

JANE   I haven't. I can look after you.

STEPHEN   That's no life for anyone.

JANE   Well it's probably my only hope of hooking a genius, so I'll chance it.

STEPHEN   I'm no genius, Jane. I can't even come up with a subject for my PhD thesis.

JANE   Perhaps you aren't searching in the right places. Try asking God that.

STEPHEN   Oh, I've tried it. I told him, now's your chance to show me you exist.

JANE   What happened?

STEPHEN   Well I did have a dream that I met Einstein, but he wasn't much help.

(*He stumbles. She holds him up.*)

You see, I can't even smooch properly.

JANE   You look after the PhD, I'll look after the
                 smooching.

                 (*They dance off. The music fades and backscreen
                 changes to show the interior of an observatory or
                 laboratory.* GOD *enters as* ALBERT EINSTEIN. *His
                 high domed and wild-haired wig makes him
                 instantly recognisable. He beams at the
                 audience.*)

GOD   Yes, it's obvious who I am now. Another god. The
                 god of twentieth century science. To progress
                 further we need a bit of serious science now.
                 Don't panic. I'll make it as easy as possible. And
                 anything you don't understand, the odds are we
                 scientists don't either.

                 (*The backscreen shows the various equations
                 connected with relativity – including* $E = mc^2$.)

                 My Theory of Relativity. What on earth does it
                 mean? $E = mc^2$. What in *heaven* does that mean?
                 It's quite simple – 'relatively'. (*Chuckles at his
                 own pun.*) Isaac Newton discovered that all bodies
                 attract each other. A force we call gravity. The
                 more massive the bodies, the bigger the attraction,
                 the further *apart* the bodies, the smaller the
                 attraction. Brilliant. I'm attracted towards all of
                 you at this precise moment, but because I'm
                 attracted a lot more strongly towards the centre of
                 the Earth I don't sense it.

                 (*The backscreen shows the solar system.*)

                 The planets in the solar system are attracted to
                 each other, but more strongly towards the more
                 massive Sun, and since the attraction is balanced
                 by their desire to keep moving in a straight line –
                 another of Newton's laws – they end up circling
                 the Sun. Excellent. Works a treat. Explains
                 practically everything. (*Pause.*) Practically . . .
                 What it doesn't take into account is what happens
                 inside those bodies.

(The diagram of the solar system merges into one of a simple helium atom, showing electrons orbiting a cluster of neutrons and protons.)

We're now into what's called quantum physics – the science of the tiny as opposed to the science of the huge. Even the ancient Greeks guessed that matter consisted of tiny particles, which they called atoms, meaning indivisible. We've had to wait a couple of thousand years to prove that atoms *themselves* are divisible – into ever smaller particles and sub-particles, all buzzing with their own methods of attraction and repulsion. A source of phenomenal energy. And that ultimately these sub-particles become so small and elemental that they are virtually nothing *but* energy. In other words – matter can become energy – and conversely energy can become matter! Mind-blowing – literally.

(Knocks his head with his knuckles.)

This is not solid. It's merely a highly complex arrangement of energy forces.

(Beams at the audience.)

And what *are* these energy forces, ultimately? Nuclear . . . electric . . . and the most conspicuous of them all – light. Let there be light.

(He measures his words.)

Therefore, since light is the essence of energy, nothing can move faster than light. And consequently the measurement of energy must be *connected* to the speed of light. Eureka – I finally worked out the correlation!

(The screen goes back to $E = mc^2$.)

The energy 'E' inherent in a body equals its mass 'm' – and here's the clever bit – times the square of 'c', which is the speed of light.

(*The formula changes to 'Energy = mass x (186,000 m.p.s.)^2'.*)

But that's a simple equation, I hear you say. What's so clever about that? Well it may seem simple, but it took me ten bloody years to work it out. Anyway, since the speed of light is a very large sum – one hundred and eighty-six thousand miles per second – you can see that matter contains a phenomenal amount of energy.

(*The screen shows a photo of an atomic explosion.*)

As the inhabitants of Nagasaki and Hiroshima discovered.

(*Screen changes to the Milky Way.*)

So how does this affect the Universe? The answer is, pretty fundamentally. Since nothing can communicate itself faster than light – everything, *including Newton's precious gravity*, is relative to the speed of light!

(*A picture appears of the Sun and revolving planets.*)

The light from the Sun takes eight minutes to reach the Earth. If the Sun was suddenly plucked from the sky, the Earth wouldn't immediately be free to fly off into space. It would have to wait eight minutes for the information to get here!

(*Beams triumphantly at the audience.*)

Which means that events are not necessarily simultaneous. Time too is relative. (*Pause.*) Time, you say? (*Nods.*) Yes.

(*Turns to the backscreen. It shows a picture of a railway carriage with a person inside.*)

Let me give you an illustration. I like this. Take someone in a train travelling at fifty miles an hour. If he throws a tennis ball forwards down the carriage also at fifty miles an hour, the tennis ball, *relative to himself*, will only be travelling at fifty miles an hour.

(*The screen illustrates the concept.*)

But, *relative to someone standing on the embankment*, the ball will be travelling at one hundred miles an hour. The speed of the train combined with the speed of the ball. If he throws it backwards, the person on the embankment will see the ball as staying in the same place. The speed of the train cancels out the speed of the ball.

(*Further illustration.*)

But now – here's the extraordinary thing – if the person on the train shone a torch forwards or backwards, the results would be quite different. Why? Because light always travels at its same fundamental speed whatever the speed of its source!

(*Further illustration.*)

The projected light is not affected by the velocity of the train – it only travels at its own precise velocity. And the chap on the bank sees it going at that speed too. It's not even affected by the speed at which the Earth is spinning. Are you with me? The speed of light is the immutable pace of basic communication within the Universe.

(*Gestures delightedly at the screen.*)

This affects the relationship between the two observers in extraordinary ways! As far as the tennis ball was concerned it took the same *time* to travel down the carriage for the fellow on the embankment as for the fellow on the train, but went a different distance at a different speed 'relative' to each of them. But speed is a ratio of distance and time: miles per hour, feet per second.

(*Simple equation on the screen.*)

If therefore you make the speed of the projected object the *same* for both of them, as with light – then plain arithmetic tells you that the *time* has to *vary* for each – relatively.

(*The equation shows the change. He beams happily.*)

Astonishing! Everyone actually carries their own personal time around with them. Which means that a clock on top of a mountain will run minutely faster than one in the valley. Oh yes, it's been proved! It means that, paradoxically, a twin who goes off on a long space journey would come back considerably younger than his twin who stayed on Earth. And the faster he goes, the slower goes his time. If his spaceship was able to travel at the speed of light itself, then his time would stop completely! He would have achieved eternal life.

(*Lines on the diagram illustrate the point. He chortles with pleasure.*)

And once you extend the idea of relativity to everything – then on the large scale of the cosmos it has astounding effects. It means that space – and time – bend.

(*Picture of Dewitt's grid of space distorted by a planet, with the curved path of a spaceship indicated round one side, and that of a light ray round the other.*)

It causes us to see stars in different positions – as well as times – from where they see themselves. Oh yes, indeed! Newton didn't catch on to that one! Turns our whole conception of things on its head, doesn't it? Better to forget the whole idea of Newton's gravitational attraction, and think instead of space and time being the vital connecting factors between objects.

(*Takes a deep breath and rubs his hands together. The backscreen reverts to the laboratory interior.*)

There you are – wasn't so bad, was it? You're all scientists now! (*Raises a finger.*) But there is one particular budding scientist whose head I need to enter.

(STEPHEN *enters wearing pyjamas.* EINSTEIN *turns to him.*)

Ah. You've got a bit of a limp there, haven't you? Sorry, all my clever theories aren't much use in your case. What do the doctors say?

STEPHEN   They're not much use either. You could be in another way, though.

EINSTEIN   What's that?

STEPHEN   My own work.

EINSTEIN   Bit stumped are you?

STEPHEN   You were once, weren't you? For years you didn't accept what your investigations were telling you – that the Universe was changing.

EINSTEIN   Don't embarrass me.

STEPHEN   Why not? Why couldn't you accept the evidence of your own discoveries?

EINSTEIN   (*angry*) Young whipper-snapper! Are you trying to insult me? Are you trying to challenge my reputation?

STEPHEN   No, no – please! I have enormous respect for your work. Without you none of modern science would be possible.

EINSTEIN   That's better. What's your point then?

STEPHEN   It's the very fact that you were so far ahead of
            everyone else that makes it so strange. That you
            couldn't accept the ultimate implication.

EINSTEIN   Well . . . (*Looks about him.*) Well, don't tell
            anyone, but I have to admit it's the most
            embarrassing fact of my life. But you have to
            remember that the whole of scientific and
            philosophical thought up to then had held that the
            Universe was a vast, fixed, eternal system. Stars
            and planets weaving around each other in a
            sublime ethereal dance. With God – (*To the
            audience.*) or somebody – (*To* STEPHEN.) as the
            choreographer. For me to accept that I personally
            had changed God's plan, and the whole caboodle
            wasn't permanent at all – might even be heading
            for the most cataclysmic blow-up imaginable – was
            . . . was . . .

            (*He is lost for words.*)

STEPHEN   Yes, I see.

EINSTEIN   I couldn't accept it for years. I even fiddled my
            own mathematics to try and make out it wasn't
            true!

STEPHEN   Yes.

EINSTEIN   But in the end Edwin Hubble – dogmatic bugger! –
            did for me. His evidence was practically thrust
            down my throat.

STEPHEN   The light spectrum from distant galaxies meant
            they were moving away from us.

EINSTEIN   (*beating his forehead with his fist*) How could I
            have been so blind? Of course the Universe
            couldn't be static! All that inherent energy that I
            had so tortuously exposed meant it had to be

expanding. Rushing outwards at a phenomenal rate.

STEPHEN   Which meant . . .

EINSTEIN   Which meant that at some stage in its past it must have been the opposite. Contracted. Condensed into . . . what? An infinite smallness? An explosive microbe? Which in turn would mean that God had not after all created an eternal miraculous heaven. But might be just blowing up a huge balloon that could go pop any moment!

STEPHEN   Or even . . .

EINSTEIN   What?

STEPHEN   That it may not be necessary to have a God at all.

EINSTEIN   Oh no! (*Out front.*) He doesn't catch me on that one. (*Back again.*) I never said that.

STEPHEN   You still believed in the biblical idea of God?

EINSTEIN   Oh, not the biblical idea, no. Like Newton, I couldn't be a scientist and believe in the literal truth of all that. But the philosophical truth – now that's different.

STEPHEN   How?

EINSTEIN   It doesn't matter how the Universe may be moving, or where it's ultimately going – it still needed something to start it off in the first place.

STEPHEN   But did it need a moral something?

EINSTEIN   What?

STEPHEN   That's the point, isn't it? My fiancée, Jane, is a firm believer in a just God. But look at the world! For what purpose does she pray to him?

EINSTEIN   You must ask her, but to me it's obvious. We have
                an emotional need for a mentor. We need to feel
                there's an ethical purpose to the Universe,
                otherwise what was the point of creating it in the
                first place?

STEPHEN   But he patently doesn't provide that, does he?
                You're a German Jew. You lived through the
                Holocaust . . .

EINSTEIN   (*sighing*) Oh dear, that old chestnut again.

STEPHEN   Well if God was a moral God, how could he allow
                such things to happen?

EINSTEIN   And if he was a moral God, how could he allow
                what has happened to you?

STEPHEN   Well, yes.

EINSTEIN   My dear boy. I gather you're quite good at the
                physics bit. Something of a whiz kid they tell me.

STEPHEN   Not yet.

EINSTEIN   Well, you really do need to extend your horizons a
                little. If we physicists are going to explore the
                ultimate regions of the cosmos then we have to go
                into other matters as well. Plato and Aristotle, the
                questions of determinism, causality . . .

STEPHEN   I haven't got time.

EINSTEIN   You've forgotten already – time is relative.

STEPHEN   Not that relative. I've got two years. I can't go at
                the speed of light.

                (*Pause.* EINSTEIN *goes to him and puts a fatherly
                arm around his shoulder.*)

EINSTEIN   Listen, dear fellow – whether you believe God has
                a hand in it or not, our duty is obvious. To make
                the best use of whatever time we have on this

particular planet – another aspect of relativity. Go and marry your lovely girl; make love to her as much as you can while you can; study whatever your instincts lead you to study. Live! And maybe you'll find you can cram more into your two years than many do in their three score and ten. (*Waves a finger.*) Relativity.

STEPHEN   (*after a moment*) Right.

(*He turns to go.*)

EINSTEIN   A tip. My work is just a half step along the way. Have a look at Darwin. He came up with the truly most important discovery so far – not me.

(STEPHEN *goes off.* GOD *takes off his wig.*)

GOD   (*examining it*) Funny how scientists can conduct the most complex experiments, yet rarely seem able to organise a simple hair-cut.

(*Out front.*)

How am I doing? Keeping my end up? It's going to get trickier though – I can feel it. That young man is a bit too bright for my liking. He unnerves me somewhat, I don't mind admitting it. Ah well – hold on to your hats!

(*The backscreen shows reams of complex mathematics. He looks at it.*)

Mathematics! The language of the Universe. Why? – I hear you ask. What is all this magic about mathematics? It's just sums!

(*Raises a finger.*)

But, you see, 'sum' is the Latin for 'I am'. Not I might be, I should be, I aspire to be. But what I demonstrably am. Mathematics is the pure, precise definition of that which – however hugely complex

– is. As well as quite a lot that isn't – but that's another matter.

(*The* Nurse *comes on with a tweed jacket and a lecturer's pointer.*)

Ah – my angel.

(*She helps him into the jacket.*)

In the 1960s possibly the most brilliant mathematician on the planet was a certain Roger Penrose. Certainly better at sums than Stephen Hawking. Who was fortunate to attend some of his lectures.

(*He squeezes the* Nurse's *bottom.*)

NURSE   Excuse me!

GOD   Just testing magnetic attraction.

(*She hands him the pointer. He watches her as she goes off, then turns to the backscreen.*)

Don't worry, I'm not going to ask you to understand that lot.

(Stephen *appears to one side, leaning on a walking stick. Listens as* God *lectures as* Roger Penrose.)

PENROSE   But what does the mathematics tell us in practice? We know that a massive body such as our own Sun is created by a collection of gas particles pulled together by their own gravity. This vast implosion creates a nuclear fusion whose energy maintains the star in its brilliant, glittering state in the heavens.

(*Diagram of a solar system in space.*)

However, eventually it burns up all its fuel, the glorious flames die, and the star contracts again like a great shrinking cinder. As our own Sun will do. Don't be too alarmed, we've got a few billion years yet. But now the real fun begins.

(*Illustration of imploding star.*)

Newton has told us that the closer the mass of particles, the greater the gravitational pull of the whole. And therefore the greater the velocity necessary for an object – whether a particle or a spaceship – to escape from it. Einstein tells us that, once the required escape velocity rises to the speed of light itself, *nothing* can escape – not even light itself.

(*Back to the maths formulae.*)

The mathematics demonstrate that it only requires a contracting star with a few times the mass of our own Sun for that ultimate escape velocity to become inevitable.

(*Pause.*)

At which point, the inward gravitational force is so immense that the star has to go on contracting until, incredibly, it reaches *zero* radius . . . zero volume. A point where its density is infinite, where even space and time are wrapped around it to . . . nothing. Or infinity – take your pick. What science calls a true singularity. Where all our familiar laws are null and void.

(*The screen clears, the lights come up.*)

Thank you.

(STEPHEN *limps forward on his stick.*)

STEPHEN   That was a fascinating lecture, Doctor Penrose.

PENROSE   Hawking, isn't it?

STEPHEN   Yes.

PENROSE   Heard about you. Very bright, they say.

STEPHEN   I was wondering . . .

PENROSE   Yes?

STEPHEN   What would happen if . . . this may sound silly . . .

PENROSE   Go on.

STEPHEN   You applied your singularity mathematics to the
             Universe as a whole.

PENROSE   (*frowning*) But the Universe is expanding.

STEPHEN   I mean the reverse process.

PENROSE   It *began* with a singularity?

STEPHEN   Yes.

PENROSE   (*after a moment*) Then what began it?

             (STEPHEN *shrugs and indicates the backscreen.*)

STEPHEN   But the mathematics might work.

PENROSE   I'd hate to try and work them out.

STEPHEN   It was just a thought.

PENROSE   (*after a moment*) I hope all your thoughts aren't
             so momentous. Let me know if you take it further.

             (*He goes.* JANE *enters. She hugs* STEPHEN.)

JANE   How about July for the wedding?

STEPHEN   If I don't get my fellowship we'll have nothing to
             get married on.

JANE   Then get it. Have you thought of a PhD subject
             yet?

STEPHEN   Yes. Just.

JANE   (*delighted*) Well then? (*Sees his dubious face.*) You don't look very happy about it.

STEPHEN   It scares me.

JANE   Marriage or the subject?

STEPHEN   Both.

JANE   Marriage isn't the end of the world, my darling.

STEPHEN   No, but the subject might be.

              (GOD *enters wearing a vicar's surplice.*)

GOD   (*out front*) I'm going to do my formal religious bit now. You see – even unbelievers need me for the really important events of life.

              (*The backscreen shows a stained glass window.* STEPHEN *and* JANE *stand before him as a* VICAR.)

VICAR   . . . and wilt thou, Jane, have this man to thy wedded husband, to live together according to God's law – if such exists – in the holy estate of matrimony? Wilt thou love him, comfort him, honour and keep him, in sickness and in health, and forsaking all other keep thee only unto him, as long as ye both shall live?

JANE   I will.

VICAR   I therefore pronounce you man and wife. You may kiss the bride.

              (STEPHEN *kisses* JANE.)

              Excellent. (*To* STEPHEN.) For a sceptic you managed that very well.

JANE   Thank you, Vicar. That was a lovely service.

VICAR   I wish you both great happiness. It's a very brave thing you're doing – in the circumstances.

STEPHEN   For Jane it is.

VICAR   For you too. Marriage is a journey as extraordinary as any trip to the stars.

(She leads STEPHEN *off.* GOD *takes off the surplice.)*

GOD   Quite a challenging situation they have there. Apart from all the usual challenges of wedlock.

(The NURSE *enters, and takes the surplice from him.)*

If she's going to aid me in my own battle for his soul, she's going to need some assistance.

(The NURSE *hands him his pointer and exits again, taking care not to turn her back on him too soon.)*

Perhaps, if Roger Penrose, the master mathematician, can lead Stephen Hawking to the end – and thereby the beginning – of the Universe, then we'll come to the ultimate question.

(The backscreen shows a quasar. He turns to it as PENROSE *again, in lecture mode.)*

PENROSE   The nineteen-sixties may be changing society for ever – they are certainly changing our understanding of science for ever. The new radio telescopes are discovering gigantic new sources of light that can only come from the outermost fringes of the Universe. These so-called quasars burn brighter than a trillion suns. They threaten to consume entire galaxies around them, and are so massive that all our Einsteinian calculations tell us each must contain a singularity at its centre. What we now call a black hole.

(*Illustration of the light spectrum.*)

Light again – that wonderful thing – tells us so much. Its brightness gives us distance, its wave frequency gives us velocity, its spectrum even gives us chemical composition. By analysing the light waves from these quasar monsters we can for the first time attempt serious calculations, not only of the size and composition of the Universe, but also the actual rate of its expansion. And by tracing this expansion backwards, we can now guess at the date of its origins in the Big Bang. Even Bible believers now have to accept that assessing the age of the Earth by adding up how many generations begat each other in the book of Genesis is no longer credible.

(*Turns as* STEPHEN *limps on, now with crutches to aid him. The backscreen reverts to mathematics.* PENROSE *comes forward to help.*)

PENROSE   Ah, Stephen . . .

STEPHEN   (*fiercely*) No! (PENROSE *withdraws.* STEPHEN *grins. His voice is becoming slurred.*) Sorry, Roger – I just feel that if I can't get across a room unaided, I haven't much hope of reaching the edge of space.

PENROSE   You're doing remarkably well. It must be at least four years since . . .

STEPHEN   Yes. I seem to have acquired a new lease of life. A combination of vitamins, work, and marriage.

PENROSE   I hear Jane is remarkable. She runs your life for you.

STEPHEN   Yes.

PENROSE   And you have a son.

STEPHEN   Not the last, I hope.

PENROSE   I'm glad that all your working parts aren't affected
               then.

               (STEPHEN *grins again*.)

STEPHEN   I need your help.

PENROSE   You do?

STEPHEN   The Americans have finally proved that microwave
               radiation is spread throughout the Universe.

PENROSE   Yes.

STEPHEN   It can only have come from the huge explosion of
               energy at the beginning of time. Everyone must
               surely accept the Big Bang theory now.

PENROSE   Most people in science do. I don't know whether
               the Church will.

STEPHEN   It fits in neatly with the biblical Creation.

PENROSE   (*beaming at the audience*) So it does.

STEPHEN   However . . .

PENROSE   What?

STEPHEN   I'm convinced I'm right. There must have been a
               singularity at the heart of it. A black hole in
               reverse. Everything . . . exploding from nothing.
               The implications would be immense.

PENROSE   Indeed.

STEPHEN   But I need your help with the maths.

PENROSE   Oh dear.

STEPHEN   I can give you the parameters, but the specifics
               are beyond me. We'd have to extend the scope of
               mathematics.

PENROSE   We'd practically have to reinvent mathematics.

STEPHEN   (*cheerfully*) That's what I like about mathematics. It's an art form really.

PENROSE   No use art without practical experiment to confirm the results. How do you empirically prove a singularity? By definition it can't be done.

STEPHEN   You worry about the maths, I'll worry about the proof.

PENROSE   Stephen – is this the right thing? It's a huge enterprise. It'll mean giving up everything else for it.

STEPHEN   I've nothing to give up.

PENROSE   Your marriage?

STEPHEN   Jane knows what I'm about. It's why she married me.

PENROSE   Your other studies. This could all come to nothing.

STEPHEN   (*grinning*) Exactly. A singularity.

(*He goes.* PENROSE *turns to the audience as* GOD *again.*)

GOD   The problem is, you see, that quantum physics – the science of subatomic particles – appears to be a totally different business from all the other sciences. If indeed it was me who set the rules, then I did so apparently in order to totally confuse mankind, which lives by a completely different set of rules. (*Beams.*) I do like to set challenges. How do I explain these inexplicable rules? Tell you what – I'll just rattle through the basics, and you see how much sense you can make of them. Or have a snooze. Humankind hasn't understood them for a quarter of a million years – it won't matter if you don't get the hang of them tonight.

(*The backscreen shows a diagram of an atom with its attendant nuclear components. He speaks fast, emphasising only the paradoxes.*)

Forget about how many angels can crowd onto the head of a pin. Where atoms are concerned we're talking billions. And to get to their component parts we have to scale down a similar quantum leap. Here particles called protons cluster together with particles called neutrons, orbited by particles called electrons. The neutrons and the protons are divided into even smaller particles called quarks – your name, not mine.

(*Diagrams of particles.*)

With these particles come so-called anti-particles which, when they make contact with the particles, cause each other to disappear. And all these types of particle are subdivided into further types labelled bosons and fermions – which are subdivided into further types labelled leptons, hadrons, photons, gluons, and gravitons – which are all too complicated to explain. Most of these particles have been around since the beginning of creation even though they're too minute to be detected – and even if you could detect them you could still never be certain exactly where they were. Because they often don't behave like particles at all, but like waves. Now these waves indicate the different forces in the Universe – such as the gravitational force, which always attracts and is the weakest force; the electro-magnetic force – which is squillions of times stronger than the gravitational force, but which both attracts and repels and so tends to cancel itself out; and two other forces – the weak nuclear force, which acts between some of the particles but not others, and can become strong – and the strong nuclear force, which also acts between some particles but not others, and can become weak. And under certain conditions the last three forces all behave like each other, and so could be called the same force.

(*Takes a deep breath and beams at the audience.*)

Got that? Like me to go through it again? I have to take my hat off to you lot – to have worked out all of that, without even being able to see most of the bits, is pretty damned clever. Especially remarkable when one considers that you're all, of course, composed of all these particles, and that collectively you behave pretty much as they do. Just a frenzied buzzing of indistinguishable specks, appearing and disappearing without notice, mutually attracting and repelling each other, serving no definable purpose, following no predictable paths, and governed by no rational system of behaviour. And progressing . . . where? To what?

(STEPHEN *and* JANE *enter, the former on his crutches.*)

Take these two. If ever there was a pair of contrasting organisms, set on different orbits, but held together by powerful indefinable forces, they are it.

(*Goes off. The backscreen shows a domestic living room.*)

STEPHEN   What's for supper?

JANE   I thought we were going out.

STEPHEN   Oh . . . I'd love to – but I've got to prepare this talk for the Royal Society. I'll need you to type it out.

JANE   (*drily*) Shepherd's pie then.

STEPHEN   Oh, good. I need a drink.

(*Hobbles to the side.*)

How are the children?

JANE   Exhausting. Robert chose a wall to fall off today.

STEPHEN   (*turning*) Weren't you watching him?

JANE   I try, Stephen, but . . . we're going to need some
            help soon, you know.

STEPHEN   No. We can manage. (*Stumbles slightly as he
            turns back.*)

JANE   Let me get it.

STEPHEN   (*going off*) I can do it.

JANE   You could let me pour you the odd drink.

STEPHEN   (*off*) I'm not helpless.

JANE   I know, but you can take this no concessions
            business too far.

            (*Crash of glass off-stage.*)

STEPHEN   (*off*) Damn!

JANE   Apart from which we'll soon have no glassware
            either. (*After a moment.*) I really think you should
            consider . . .

STEPHEN   (*appearing again with a glass*) What?

JANE   A wheelchair. Just for general getting about.

STEPHEN   (*irritably*) I don't need one!

JANE   You'd save yourself so much energy.

STEPHEN   I've got lots of energy.

JANE   I know, my darling, but you need all you have for
            your work.

STEPHEN I can't give in to it, Jane. Every day I deny it is a day longer than the doctors gave me.

JANE Might it not be . . . that God is granting them?

STEPHEN (*sighing*) If God has chosen to grant me absolution, then why did he condemn me in the first place?

JANE Perhaps to see how you would cope with it.

STEPHEN Oh, I see – so having watched me struggle through the course, he now gives me a prize for effort.

JANE (*with feeling*) I wish you wouldn't be so cynical.

STEPHEN But you see, Jane, when one is trying to interpret a universe that's developed over aeons of trial and error, it's rather hard to accept the idea of a celestial headmaster organising it all.

JANE It's a better idea than assuming we're at the mercy of any old cosmic bang that might come along! It gives me comfort at any rate.

STEPHEN Yes – that's what it's all about.

JANE What?

STEPHEN Comfort. People need comfort in such a bewildering world. That's why so many otherwise intelligent beings cling to the concept of a guiding force. Unfortunately it's not a very logical concept.

JANE Oh, the hell with logic!

(*Storms off.* STEPHEN *downs his drink and follows.* GOD *comes on from the other side, as the backscreen shows Michelangelo's Jehovah from the Sistine Chapel paintings, or some such.*)

GOD   I don't think I like this headmaster metaphor much either. Is that how you lot see me? Judging from most of what you say about me it probably is. 'The wrath of God!' 'God the chastiser.' Please! Grant me a little more discrimination than that. (*Nods after* STEPHEN.) Even if he and most of his fellow scientists won't grant me anything. But that, you see, is because they're winning at the moment. They're on a high. The human race has never seen such an era of pell-mell scientific advances. With him in the vanguard. The world of science beating a path to his door more than a decade after the pronouncement of his death sentence. I told you he was going to give me trouble.

(*Stands on one side.* STEPHEN *comes on wheeling himself in a wheelchair. Comes to the centre forestage and stops. The backscreen shows a view of a beautiful spiral galaxy. His voice is now quite slurred but distinguishable.*)

STEPHEN   (*referring to the backscreen*) The spiral galaxy M100 in Virgo, fifty million light years away. Beautiful, isn't it? Over and over again, science makes us redefine our philosophy. Einstein's discovery of relativity proved that our Newtonian vision of a fixed unchanging cosmos was no longer valid. Everything is relative, everything is variable, including what we thought was most immutable – time and space.

(*The backscreen shows a supernova.*)

Then Edwin Hubble proved that not only is the Universe idiosyncratic, but also it is *expanding* at a phenomenal rate – and presumably has done so since its beginning. Which means that it had a beginning – somewhere, somehow.

(*The backscreen shows a huge cosmic explosion.*)

This gave comfort to those who believed in a Creator, whilst undermining the belief that he has since taken any part in organising his creation.

(*Pause. The backscreen shows maths calculations.*)

Now Roger Penrose and I have proved that this beginning, this so-called Big Bang, must have originated out of a singularity. A void which held within its zero confines the capacity for the gigantic explosion of energy from which everything – including space and time – was born. How did this extraordinary event occur? Amazing though it may seem, I believe it is within the power of science to discover this. We are not so very far from analysing that moment of creation. And were we to do so, it would then seem that there is no need even for a Creator.

(*He wheels away.* GOD *stirs himself.*)

GOD   Huh! We'll see about that. I told you the scientists were getting cocky. They're having a hot period. But it won't last. There's another law of the Universe, you see, that they've forgotten. The Pendulum Law. The greater the movement in one direction, the greater the reaction in the other. First the Big Bang – then the Big Deflation. Study your history – it always happens. Whether to empires, philosophies or theories. I can wait.

(*Flicks his fingers at the backscreen. It changes to William Blake's 'God Creating the Universe'. He smiles at the audience and starts to go. Then stops.*)

However, a word of warning. Despite the fact that this is me speaking – those of you who take strength from the idea of virgin births and miraculous resurrections, those of you who think salvation is a matter of not eating pork, or of placing your prayer mat correctly every few hours, don't look for reassurance here. Hold to these notions by all means if you gain comfort from them. But we are dealing now, as is surely clear, with matters that go far beyond such earth-bound conceptions. With numberless light years of distance, with time beyond measure, with unquantifiable legions of particles, and planets encompassing forms of life that only I can imagine, and amongst which the human species is but a flick of my finger.

(*Flicks his fingers. The backscreen changes to a vision of hell by Hieronymus Bosch, or similar.*)

If you want to come further with me you must move beyond your superstitions and your mystical yearnings. You must grasp your courage, summon your intellectual vigour, and take the leap with me to a far grander vision.

(*Pause.*)

If, that is, I exist.

(*Smiles again and exits.*)

Blackout.

ACT TWO

The backscreen shows one of Salvador Dali's distorted time paintings, or similar.

GOD enters, wearing normal clothes, and comes forward to the centre stage.

GOD   Such mystery to life. Time varies. Space bends. Light vanishes into dark holes. You may ask, is any of this relative to me? The time I have to get to work doesn't vary. The space in my bank balance doesn't bend. Not much light at the end of my dark tunnel.

(Picture of the cosmos.)

Do you have any part to play in this infinity of space and time, or are you just another invisible particle, flickering for a moment, riding the microscopic wave, and then – pht! Like Schrödinger's rabbit, lived and died in the same instant?

(Raises a finger.)

Or is it conceivable that your ephemeral existence has some effect in the whole vast paraphernalia? That you too carry a vital charge that helps to motivate the cosmos?

(STEPHEN enters in his motorised wheelchair and moves in an arc round the stage.)

Think of it this way. Perhaps he's a prototype for the future of humanity. An increasingly inventive brain in an increasingly irrelevant body. Mind over matter. Could thought extend, as well as interpret life?

(STEPHEN is going too quickly and nearly hits an obstacle.)

STEPHEN   Bugger!

GOD   Perhaps not yet. Let's not get too carried away.
             1979 – more than fifteen years after his
             pronounced death sentence. He is thirty-seven,
             Fellow of The Royal Society, and the recipient of
             honours from around the world. He has even had
             the ultimate accolade – his name attached to a
             scientific phenomenon – Hawking Radiation. The
             discovery that black holes aren't quite black after
             all, but radiate phantom particles. Don't even
             attempt to understand. But it caused a huge furore
             amongst the boffins, and now he receives the
             highest reward of his career – Lucasian
             Professorship of Mathematics at Cambridge. The
             post once held by Isaac Newton himself. That
             deserves a party to celebrate.

             (*Music and flickering disco lights.* GOD *takes
             spectacles from his pocket and puts them on.*)

             I am now going to gate-crash the party as
             someone who is to have a radical effect on the
             Hawking household.

             (JANE *and the* NURSE *come on and dance, as*
             STEPHEN *whirls about in his wheel-chair.* GOD
             joins in. They all revolve around STEPHEN *like
             planets orbiting a sun. The music stops. Everyone
             laughs and claps.*)

STEPHEN   (*slurred*) I am so grateful to everyone who over
             the years has made this possible. My supervisors,
             my colleagues, my students . . .

JANE   (*dry*) Your wife.

STEPHEN   (*nodding*) And my wife. Without her I would
             probably not be alive. And I *certainly* wouldn't be
             where I am.

             (*More applause.* JANE *kisses him. The* NURSE *helps
             him to a drink.* GOD *turns out front.*)

GOD Very true. I know. I am one Jonathan Hellyer Jones – musician, choirmaster . . . and devout believer in me!

(*He moves to one side with* JANE.)

JONATHAN (*gentle and unassuming*) A nice tribute.

JANE In public. It's different in private.

JONATHAN Difficult times?

JANE Oh, Jonathan, if only you knew.

JONATHAN He gets all the accolades, you get all the responsibility.

JANE Sometimes I'm so exhausted I just want to climb into one of his black holes and disappear.

JONATHAN I can imagine. Running his meteoric life as well as being mother to three children . . .

JANE That's just it. I'm not just mother to three children – I'm mother and father to four.

(STEPHEN *wheels over to them. The* NURSE *goes.*)

STEPHEN Are you flirting with my wife?

JONATHAN You could say that. I'm wooing her to join the church choir.

STEPHEN Choir?

JONATHAN She has a lovely voice, Jane.

STEPHEN (*surprised*) Has she? I didn't know.

JONATHAN Untrained as yet. It would give her an interest of her own.

STEPHEN   Oh, she'd be interested in anything to do with the church. You'll . . . be seeing a lot more of her then?

JONATHAN   You don't have to worry, Stephen. We will keep the strictest boundaries.

STEPHEN   Hah – that's ironic.

JONATHAN   What is?

STEPHEN   I'm working at this moment on something I call the no-boundaries theory.

JONATHAN   What's that?

JANE   Don't let him start!

JONATHAN   No, I'm truly interested.

STEPHEN   (*pleased*) Are you? Well, science more or less accepts now that the Universe began with the Big Bang – where mass and time and space were all compressed to nothing.

JONATHAN   Nothing? (STEPHEN *nods*.) Can it be proved?

STEPHEN   That's where the mathematics take over.

JANE   (*impatiently*) You see – in the end everything comes down to mathematics!

STEPHEN   It is hard to explain to a non-scientist, but maths is the grammar of the Universe. Totally scientific, utterly beautiful.

JONATHAN   I can understand. Music is the same.

STEPHEN   (*triumphant*) Yes! Well mathematics has told us what the exact physical state of the Universe was one trillionth of a second after the Big Bang banged. When it was still the size of a pin-point.

JONATHAN   But that is still a boundary. And while it's there –
            even if it's only a trillionth of a second – don't
            you still have to ask, ah, but who *started* the Big
            Bang?

JANE   Exactly!

JONATHAN   Who created the something out of nothing?

STEPHEN   Aha – but that's the point of the no-boundary
            theory. It says that after all there *wasn't* just
            nothing at the start of the Universe.

JONATHAN   What was there?

STEPHEN   The end of the Universe.

JONATHAN   (*puzzled*) The end?

STEPHEN   (*enthusiastic*) 'In my end is my beginning,' as
            your Bible says! Time bends, we know. It's
            inextricably linked to space. At the speed of light
            it ceases to exist altogether. Suppose space-time
            eventually bends so far it comes back on itself?

JONATHAN   You've lost me.

            (STEPHEN *circles them both in his wheel-chair.*)

STEPHEN   Imagine a piece of music – a rondo – that goes
            round in a circle, to end as it began. Yet you can
            keep the whole development of the piece in your
            head. Now imagine the life of the Universe is like
            that – a self-contained dimension. No end and no
            beginning – just an eternal circle.

JONATHAN   The rondo still has to have a composer.

STEPHEN   Ah, that's the wonder of it. The Universe is it's
            own composer.

JANE   You can't argue with him, Jonathan.

STEPHEN   You can argue all night. I love arguments.

JONATHAN   (*smiling*) Another time.

JANE   (*taking* JONATHAN'S *arm*) He still hasn't proved it yet. And what's more, he's been invited to a scientific conference at the Vatican – to explain it to the Pope. Now that I can't wait to hear!

STEPHEN   I like the choir idea. Teach her how to sing.

                (*He wheels to the front stage.* JANE *and* JONATHAN *go off. A* Dies Irae *plays, and the backscreen shows 'The Transfiguration' by Raphael or similar.* STEPHEN'S *voice is now amplified with a slight echo.*)

STEPHEN   . . . and so we now reassessing the classical conception of the Universe. We now see it as commencing with a gigantic explosion of energy some time in the past, and expanding ever since – the galaxies forming out of the vast turmoil of that expansion. (*Pause.*) In which case, two possibilities. Either it might go on expanding for ever – the matter within it gradually cooling and decaying, until ultimately it is just a formless mass of energy once more. Or alternatively the gravitational attraction between all the galaxies might gradually slow down its expansion, until it eventually stops. At which point it would then start to contract again, and hurtle in towards . . . what? A Big Crunch? Blackout? The end of creation?

                (*The backscreen now shows two spheres side by side. One is headed 'the Earth', indicating north and south poles and equator. The other is identical but instead is headed 'the Universe', and indicates 'Big Bang' and 'Big Crunch' at the poles, and 'maximum expansion' at the equator. Stephen reverses his chair to look at the screen.*)

                Either way we have a problem. Is the Universe to end again – back to an infinite obscurity or to nothing at all? Unthinkable, surely. Or could there be an alternative? Our dilemma as living beings is that we are so conditioned to our idea of linear time. Our mortal selves moving inexorably from past birth to future death. But time, we now know, is an essential commodity of space and matter – created purely for their benefit. At the speed of energy, time ceases to tick. So suppose we think of the history of the Universe as a sphere – just like the Earth, but with the extra dimension of *time* added. If the three dimensions of space and matter expand or contract from nothing but energy *to* nothing but energy, then perhaps the fourth dimension of time does so along with them. Hard for the mind to grasp, but if we could stand outside in a *separate* dimension, we could see it all!

(*Reverses back to face the front.*)

Thank you for being so attentive.

(*The sound of applause. The backscreen shows a background of Vatican architecture.* JANE *enters and stands nervously beside* STEPHEN. GOD *enters dressed as Pope John Paul II. He addresses the audience.*)

GOD   You can't say I don't give myself suitable parts.

    (*Goes to the pair, smiles at* JANE, *and kneels down to* STEPHEN'S *level.*)

STEPHEN (*protesting*) Please, Your Holiness . . .

POPE   (*waving away his protest*) You can't kneel to me, so I shall kneel to you. My people tell me you gave an interesting talk, my son.

STEPHEN Thank you, Holiness.

POPE   You have done remarkable things with your life. God has made you both an example for all who are afflicted.

(STEPHEN *nods*.)

Now, they say your work on the origins of the Universe is greatly important. That is good. The Church no longer condemns the explorations of science, as it has done, sadly, in the past. (STEPHEN *nods again*.) And this Big Bang you call it? – yes, that sounds very much how God would have chosen to commence his creation.

(*Lays a hand on* STEPHEN's *knee*.)

But remember, my son – you should not presume to stray too far beyond the first moment. That is God's province, not man's. Man must remain humble when he reaches God's . . .

STEPHEN   Boundaries?

POPE   (*nodding*) Boundaries.

STEPHEN   Yes, Your Holiness.

POPE   They tell me you are not a strict believer in Jesus Christ?

STEPHEN   (*awkward*) Science finds it hard to accept the literal truth of miracles, Holiness.

POPE   Despite the miracles out there?

STEPHEN   Not miracles. Provable phenomena.

POPE   Remember Christ's forty days in the wilderness. He too had to ponder the mysteries of existence. They are in our heads as well as out there.

STEPHEN   That I do believe.

POPE   (*rising*) So if God did not create the power of thought, who did?

(JANE *nods approvingly. He points at* STEPHEN's *head*.)

The more extraordinary grow the workings of the brain, the more significant becomes that question.

STEPHEN   Or, Holiness, the more evident become the workings of evolution.

POPE   Ah.

(*The* POPE *nods, glances at* JANE, *and leaves.* JANE *gasps in frustration.*)

JANE   Oh! He couldn't have known what your talk meant! The Pope would never have accepted such a theory.

STEPHEN   I think it was rather above the heads of his minions. They didn't understand the implications.

JANE   It's above my head, Stephen, but I understand the implications. No beginning to the Universe, so no God to begin it.

STEPHEN   Not in the sense you mean, no.

JANE   Look around you. St Peter's, the Sistine Chapel – the most wonderful buildings in the world – all built in honour of God. Are they a delusion?

STEPHEN   Versailles, the pyramids, the Taj Mahal – all built in honour of man. It's his invention which is the wonder.

JANE   Oh, there's no arguing with you. I've had enough of wondrous buildings. Let's go home to our little house.

(*They move across the stage as the backscreen changes to show their domestic scene.*)

Tell me this. Can your beloved science explain the composition of the flowers in our garden?

STEPHEN   You wouldn't deny the scientific explanation of a rainbow. Likewise a flower has the most efficient, and therefore beautiful, structure for its evolved lifestyle.

JANE   But you can't reduce beauty to mathematics!

STEPHEN   You can. As Jonathan recognised, even a Mozart symphony may be just a harmonious conjunction of universal vibrations and sound waves.

JANE   But our response to things! Our deep, individual response . . .

STEPHEN   Is our primeval response to the environment. It's the butterfly knowing instinctively that this is the right day, the right leaf. The deer thinking, yes, this is the breeding ground for me! Humanity has just refined it far beyond its purpose – as humanity refines everything.

JANE   Well, it hasn't refined things too well round here.

STEPHEN   What?

JANE   This breeding ground isn't functioning very well at the moment. I'm exhausted, Stephen. I'm not coping very well.

STEPHEN   You're doing fine.

JANE   (*passionate*) I'm not! I'm not doing fine. Oh, I want to manage everything for you, but I . . . people don't realise – you're a whole industry in yourself. The Earth is just a microscopic speck to you, but getting you round it time and again is such a huge operation. Our children don't conform to the straightforward laws of physics, this household doesn't run on atomic energy, all the people who want your attention don't communicate via . . . cosmic radiation! (*She is close to tears.*)

STEPHEN   Come here.

(*She goes to him and he holds out a hand clumsily.*)

JANE   It's not just that. I feel I've lost my own identity in the shadow of yours. This is one butterfly that . . . can't find its leaf.

STEPHEN   You don't need a leaf. Our realm is the Universe.

JANE   It may be for you, Stephen. It's not enough for me.

STEPHEN   (*ironic*) Not enough?

JANE   The Universe gives you your purpose. I don't know what the purpose of my Universe *is* – beyond the day-to-day struggle. Which you don't have to deal with.

(STEPHEN *just looks at her.*)

I'm sorry. That sounded cruel. I know what you have to deal with.

STEPHEN   I realise that in one sense it leaves me free.

JANE   Yes.

STEPHEN   That also leaves me with a responsibility.

JANE   To what?

STEPHEN   Justify my freedom with results.

JANE   But you can't go to the ultimate end, Stephen. *You* aren't God.

STEPHEN   (*after a moment*) Who is this God you believe in so passionately?

JANE   If I could answer that I wouldn't have to argue with you about him.

STEPHEN   Then how do you know he exists? Tell me – how?

JANE   Faith.

STEPHEN   Faith decreed the Earth was at the centre of the
                solar system, and they burned men alive for
                saying otherwise. Faith has run red throughout
                the Inquisition and the Middle East and Northern
                Ireland. We need facts, not faith.

                (*She stares at him.*)

JANE   You see? I want to discuss domestic
                arrangements, and here we are talking philosophy
                again.

STEPHEN   Can you discuss domestic arrangements with
                Jonathan?

JANE   His wife died of leukaemia. He understands what it
                is to look after an invalid.

                (*She starts to go, then turns.*)

                If you had the chance of gaining your health and
                losing your science, would you take it?

STEPHEN   (*after a moment*) No.

                (*She turns and goes. He sits in silence.*)

                (*The backscreen shows an interior of Buckingham
                Palace.* GOD *enters dressed as the* QUEEN *– tiara,
                spectacles and all. Waits until the inevitable
                laughter subsides.*)

GOD   (*sternly*) Don't laugh. I am the anointed head of
                my own Church in England. Defender of the Faith.
                And I should get an Oscar for this.

                (*Goes to* STEPHEN.)

QUEEN   Professor Hawking.

STEPHEN   Your Majesty. Forgive me for not getting up.

QUEEN   Of course.

               (*She hangs an insignia round his neck.*)

               Your CBE. Congratulations.

STEPHEN   Thank you, Your Majesty. It is a great honour.

QUEEN   Just one among many, I gather.

STEPHEN   I have been very fortunate, ma'am.

QUEEN   Perhaps will power has had something to do with
               it.

STEPHEN   Pig-headedness, my wife would say.

QUEEN   Ah well – we have a little of that in my family. Tell
               me, Professor Hawking – I'm intrigued. Will
               scientific advances really mean the end of religion,
               do you think?

STEPHEN   (*carefully*) I think, ma'am, as long as there are
               always things we don't fully understand, there will
               always be belief in a power greater than our own.

QUEEN   But not one you share?

STEPHEN   I . . . I believe in an influence, if you like, that
               shapes our destiny.

QUEEN   An influence?

STEPHEN   Bernard Shaw talked of a life force. The Universe
               certainly has that.

QUEEN   I am quite a traditionalist in these things, you see.
               One would feel disturbed if all one's beliefs were
               to be seriously threatened by the march of
               technology. They are under enough threat as it is.

STEPHEN   However fast science advances, ma'am, people will
               always feel the need to find the moral way.

QUEEN   They seem to be finding the search very difficult.
            Especially in this era.

STEPHEN   Both searches are difficult, ma'am. Searches
            without end, you might say.

QUEEN   But you have said there may be an end to science.

STEPHEN   (*caught*) Er . . . only in so far as our own abilities
            limit us. Just as our ability to know God is limited.

QUEEN   But you do believe in a God?

STEPHEN   I . . .

QUEEN   Yes?

STEPHEN   Believe.

QUEEN   Yes. I see. Right.

            (*Looks at his wheelchair.*)

            Well technology seems to be helping you at any
            rate. Can you go anywhere in that chair?

STEPHEN   Almost, ma'am.

QUEEN   I'd quite like one of those myself. Help me get
            round this huge place. Goodbye, Professor.

STEPHEN   Goodbye, Your Majesty.

QUEEN   (*stopping*) One other thing. This uncertainty
            principle. One keeps hearing about it but one is
            never quite certain what it means.

STEPHEN   It means that, the more precisely we try to pin
            down the action of an elementary particle, the less
            precisely it appears to behave.

QUEEN   (*thoughtfully*) Rather like my never knowing quite what any one of my subjects is up to at any given moment.

STEPHEN   Something like that.

QUEEN   Yes, I think I understand now. One lives and learns.

              (STEPHEN *wheels to one side.* GOD *takes off his royal spectacles.*)

GOD   He handled that pretty craftily, I thought. At least he didn't have the nerve to argue with the accepted head of the Church. But even there, you see, I have a problem. The monarch has to profess faith in the one true Christian God, whilst presiding over a population that worships a dozen others. Now *I* may know they are all just different versions of me, but your history would have been a lot less bloody if humanity had realised it!

              (*Goes. The backscreen changes to the Hawking household.* JANE *enters.*)

JANE   Congratulations.

STEPHEN   Another bauble. They become meaningless after a time.

JANE   (*going to him*) No they don't.

              (*She straightens the insignia on his chest.*)

              They make you very proud. And me.

STEPHEN   (*studying her*) You look happier.

              (*She kneels beside his wheelchair.*)

JANE   Stephen, I want you to consider something. All these years I've managed on my own. I've denied my own ambitions, my own career, my needs as a woman . . . because your needs have been more

important. I make no complaint. It was the bargain
I made when I married you. But I am near the end
of my resources. And now . . . God, or chance, or
whatever you like to call it, has offered me a
lifeline.

STEPHEN   A lifeline?

JANE   Jonathan, through the goodness of his soul, and
his fondness for us all, has suggested that he
devotes himself – in the absence of any family of
his own – to ours.

STEPHEN   Devotes himself?

JANE   To me – you – the children. He wants to provide
the strength that you've never had to give. The
commitment you have not the time for. Without
threatening our marriage in any way, he wants to
. . . fill the void.

(*Pause.*)

STEPHEN   Why?

JANE   For my sake. I'm in desperate need, Stephen.

STEPHEN   But why would he do it?

JANE   Because he loves me.

STEPHEN   He loves you?

JANE   Yes.

(*Pause. Then* STEPHEN *finally nods.*)

STEPHEN   I see.

JANE   But he's a Christian. He will obey the rules.

STEPHEN   Will you?

JANE   I will try – as always.

(Jane *kisses* Stephen's *forehead, then turns to go*.)

STEPHEN   You understand, I . . .

(*She turns back.*)

I have to be this way. I have to be so single-minded. I cannot afford to deviate.

JANE   I understand.

(*She goes. The backscreen changes to* Stephen's *office. He produces some papers from his wheelchair pocket and shouts.*)

STEPHEN   Ann!

(*The* Nurse *comes on dressed as a secretary.* Stephen *waves the papers tetchily.*)

Ann, the Second Law of Thermodynamics is one of the fundamental laws of science. Do you know what it states?

NURSE   Not really, Professor Hawking.

STEPHEN   It states that the degree of disorder within the Universe must always increase.

NURSE   Oh.

STEPHEN   However, that law does not apply in this office. Will you please take this away and see that it is typed correctly and put in the *proper* order.

NURSE   (*almost in tears*) Yes, Professor.

(*Starts to go, then stops.*)

Does that law mean then that the whole Universe is going to end in chaos?

STEPHEN   Could be.

NURSE   (*sniffing*) Not much point in us all trying then, is
             there?

STEPHEN   Don't worry, you've still got time to get that done.

             (*She stifles her tears and goes.* GOD *comes on
             dressed in trousers and carrying a jacket.*)

GOD   You know, he really is starting to behave like me
             himself! It's time he got his head out of the clouds
             and back to more mundane matters.

             (*He puts on the jacket.*)

             Such as dealt with by the Administrator of the
             Cambridge Institute of Astronomy. A fixed and
             solid body amongst a galaxy of shooting stars.

             (*Goes to* STEPHEN.)

SIMON   Bullying my secretaries again, Stephen?

STEPHEN   Simon! Why can I never get anything done
             properly round this department?

SIMON   Perhaps you're asking them to do too much.

STEPHEN   Are they complaining?

SIMON   No, Stephen. They all get a buzz out of working
             here. But you should remember they're human –
             unlike yourself.

STEPHEN   (*sighing*) I'm sorry. I just . . . get frustrated
             sometimes.

SIMON   We all understand that.

             (*Crouches at* STEPHEN'S *side.*)

             Do you get frustrated in other ways?

STEPHEN   (*dry*) Simon – what are you suggesting?

SIMON   Not that. I mean, you've been at Cambridge for a long time now.

STEPHEN   I like Cambridge!

SIMON   Yes, but we're aware of the sacrifices you're making by staying here. Cambridge can't provide the sort of salaries or facilities America could offer you.

STEPHEN   The only facilities I need are in my head.

SIMON   Even so . . . your head can't conjure up funds from nowhere.

STEPHEN   You've been talking to Jane.

SIMON   You're rushing around the world. You're needing ever more professional nursing care. You're trying to educate three bright children . . .

STEPHEN   We manage.

SIMON   It's going to get worse. Can you manage in the future?

STEPHEN   (*grinning*) The future's my subject – I'm working on it.

SIMON   Don't joke, Stephen.

STEPHEN   What do you suggest?

SIMON   Your scientific books have done well, but they're for scientists. The ordinary public can't make head or tail of them. And yet the ordinary public are dying to know what it's all about.

STEPHEN   Are they – *really*?

SIMON   Yes! They hear all these extraordinary stories about black holes, big bangs, no boundaries, but they haven't the foggiest what it all actually means.

STEPHEN   So?

SIMON   Write a book for them. Write a book that puts it all in layman's language.

STEPHEN   You *have* been talking to Jane.

SIMON   She suggested it, yes.

STEPHEN   (*terse*) Impossible. It's too complex.

SIMON   No, it isn't. Just keep the mathematics out of it. Think simple.

STEPHEN   That's an oxymoron.

SIMON   There's money to be made there, Stephen.

STEPHEN   (*sighing*) I'll give it some simple thought.

SIMON   But remember – no equations.

STEPHEN   No equations!

SIMON   To the layman they're just sums. Every sum you put in will halve your sales.

STEPHEN   That's an equation! You can't explain life without equations.

SIMON   God manages. The Bible sells all right.

STEPHEN   The Bible's a parable. It's not about science.

SIMON   Nevertheless, if you want to spread *your* message, no sums!

STEPHEN   No sums.

(STEPHEN *wheels to one side. The* NURSE *enters again as the secretary. She holds out the papers to him once more.* GOD *turns to the audience.*)

GOD You might wonder why I'm trying to help promote the opposition. I work on the theory, the better the devil you know. I can't after all disown my own creation – if I created it. Therefore the more the products of my creation understand that creation, the more they will ultimately understand me. If I'm here. Hey ho! The riddles don't get easier.

(*Turns to look at* STEPHEN. *The latter nods to the* NURSE *and she takes the papers off.*)

Where to now? I think it's time we brought some *soul* into the proceedings.

(*Moves to one side, flicking his finger at the backscreen. It shows another magnificent stained glass window.* JANE *enters wearing a white surplice. Organ music sounds – the introduction to a vocal solo (such as the Bach/Gounod 'Ave Maria' or Mozart's 'Laudate Dominum', or 'I Know That My Redeemer Liveth'.). She sings the whole anthem in a pure, high soprano, with organ accompaniment. The last notes fade away. She goes off. Pause.* GOD *turns out front.*)

Well if that doesn't prove my existence, what will? And might it give me the chance to get a bit closer to him, I wonder?

(*Turns to* STEPHEN. *Coughs discreetly.* STEPHEN *turns his head.*)

She has a lovely voice, your wife.

STEPHEN Yes.

GOD Can your science *really* explain that?

STEPHEN (*taken aback*) Sorry?

GOD   Science . . . art.

STEPHEN   (*looking after* JANE) I don't know. Can God?

GOD   God's words.

STEPHEN   No. Man's words. And music. They're just *about* God.

GOD   But where did the inspiration spring from?

STEPHEN   (*looking at him curiously*) Who are you?

GOD   Oh . . . just a visitor.

STEPHEN   Ah.

GOD   You haven't answered my question.

STEPHEN   Man is a remarkably evolved creature.

GOD   Evolved, ah. You concur with Darwinian theory then?

STEPHEN   Who doesn't? Except for a few reactionary eccentrics. Don't you?

GOD   Oh yes, yes . . .

STEPHEN   I had a dream about Einstein once. In it he said that Darwin's discovery was more important than his. I think he was right.

GOD   Why?

STEPHEN   Einstein was just an important sign-post along the way. Darwin is the whole journey.

GOD   The *whole* journey?

STEPHEN   The system by which life develops. Survival of the fittest. Trial and error. It's a remarkable method.

GOD   (*disgusted*) Trial and error!

STEPHEN   You find the idea distasteful?

GOD   It's pretty . . . haphazard.

STEPHEN   But so logical. One competing life form survives better than another, so it passes on its attributes to its successors. Fail, and its failings die with it. A far more credible method of progress than some celestial autocrat tinkering about and not doing it very well.

GOD   Steady now!

STEPHEN   You must be a man of the church. Don't take it so personally.

GOD   (*recovering his composure*) Tell me then – you think life *does* progress?

STEPHEN   Without question. Look at humanity. A few hitches along the way, but you can't deny we're getting better. Well, most of us.

GOD   Pretty big hitches, some of them. Can you call the dreadful trials of the last century progress?

STEPHEN   The greater the blunder, the greater the reaction against it – that's what trial and error means! Several million people die and for the first time in history we recognise war as an evil, not a glory.

GOD   A terrible price.

STEPHEN   It's a terrible journey. But think – slavery, torture, ethnic cleansing are already things of the past for most civilised societies.

GOD   What an optimist you are.

STEPHEN   Ah! You're a politician!

GOD   Lord, no! Politicians all want to be God.

STEPHEN   You see, there's no need for a God. That's what's
            so startling about evolution. The Universe is its
            own motivator.

GOD   Ah, so it's trial and error for the rest of the
            Universe too?

STEPHEN   Of course! All that energy fighting for a way to
            express itself. All those quantum particles
            competing to fuse in the most dynamic
            combinations.

GOD   And likewise all those stars up there competing
            for *their* place in the grand design?

STEPHEN   No! Competing to *create* the grand design. It is
            happening by experiment!

GOD   Experiment – hmm. Well, most interesting
            conversation.

            (*Offers to shake hands.*)

            I'm so pleased we've met at last.

STEPHEN   I'm sorry – I can't shake hands very well.

GOD   Ah. (*Touches* STEPHEN's *hand with his own.*) God
            be with you anyway.

STEPHEN   Who did you say you were?

GOD   Let's just say, I'm an admirer.

STEPHEN   Oh. Thank you.

GOD   (*aside to the audience as he goes*) Of my own
            handiwork. Experiment indeed!

            (*The backscreen shows the Hawking residence.*
            STEPHEN *wheels slowly across the stage. The*
            NURSE *enters and comes to him with a mug.*)

NURSE   Tea?

STEPHEN   (*nodding*) Thank you.

(*She helps him to drink. He is very disabled now.*)

NURSE   Toilet?

(*He shakes his head.*)

Move a bit?

(*She gently helps him to shift in his chair, rearranges his cushion, etc. GOD as SIMON comes on, and stands watching. STEPHEN looks up.*)

STEPHEN   Simon. (*Pause.*) Are you very angry with me?

SIMON   About the book? No.

(*The NURSE goes off.*)

STEPHEN   I had to go with the Americans, Simon. They offered me more than Cambridge could ever afford.

SIMON   I realise that.

STEPHEN   (*after a beat*) You were right. I don't know how long I've got left, you see. I have to provide for Jane and the kids.

SIMON   But don't you think you should ease up, Stephen.

STEPHEN   Ease up?

SIMON   Rushing off to every conference that's going. Symposia in every corner of the globe.

STEPHEN   I have to get in as much as I can, while I can, Simon. We're so near. So close to the final breakthrough. We could have the essential answers to the whole of creation within a matter of . . .

SIMON   What? How long?

STEPHEN   A decade or two.

SIMON   (*sceptical*) Oh?

STEPHEN   Perhaps a generation or two. But a mere spark in
            the . . . hah! – brief history of time. (*Grins.*)

SIMON   Is that the name of the book?

STEPHEN   Like it?

SIMON   (*nodding*) It's good.

STEPHEN   And I'm right in the middle of all of it. You've no
            idea what it feels like. I just . . .

SIMON   What?

STEPHEN   Think sometimes I'm not quite going to make it.

            (SIMON *puts his hand on* STEPHEN's *shoulder.*)

SIMON   We none of us quite make it, Stephen. Getting
            close is probably the best mankind can hope for.
            You've got pretty damned close.

STEPHEN   I want to get closer.

            (SIMON *smiles and leaves him.*)

GOD   (*exiting, to the audience*) Nearer, my God, to thee.

            (STEPHEN *sits alone again. Then suddenly his
            throat contorts and he fights for breath. His face
            gets purple with the effort of breathing. He
            struggles in his wheelchair, then starts to slump.
            The* NURSE *runs on. She quickly assesses him, then
            rushes him and his wheelchair off-stage. The
            backscreen shows the bright windows of a
            hospital ward.* JANE *runs on from one side, as* GOD
            *enters from the other wearing his white doctor's
            coat.*)

DOCTOR   Mrs Hawking?

JANE   How is he, Doctor?

DOCTOR   It's quite serious, I'm afraid. He has suffered a failure of his respiratory system. We have him on life support.

JANE   What . . . ? What will happen to him?

DOCTOR   Well, we fear he may have contracted pneumonia.

JANE   Oh, God . . .

DOCTOR   Hopefully I can get him through that. But there is another problem. I don't think he'll be able to breathe normally again. If we take him off the ventilator he'll suffocate.

JANE   Then what . . . ?

DOCTOR   The only alternative is a tracheotomy.

JANE   What is that?

DOCTOR   It means cutting into the windpipe and implanting a breathing device – here.

(Indicates his neck just below his Adam's apple.)

That will enable him to breathe by himself in the future, but . . .

JANE   But?

DOCTOR   He would probably never be able to speak again.

(Silence as JANE stares at him.)

It's his only chance, Mrs Hawking. Quite frankly it's astounding that he's managed to live all these years as it is – let alone do all the things he's done. This might – *might* give him a little time yet.

(*The* NURSE *wheels* STEPHEN *on in his chair, covered. The bleep of a life support system is heard. He is unconscious. The* DOCTOR *and the* NURSE *leave.* JANE *goes and kneels by the chair. A moment of silence.*)

JANE   Oh, Stephen . . . will you forgive me if I let them do this to you? Will you still find a way to talk to us about the mysteries out there? Will you still be able to argue with me about God?

(*Rises.*)

Whether you like it or not, I'm going to pray to him for you.

(JANE *goes.* GOD *enters from the other side and looks after her.*)

GOD   I'm glad there's someone left who will still speak to me.

(*Turns to the audience, as the* NURSE *enters and wheels the chair away again.*)

But what can I do? It must be pretty obvious by now that, if indeed I did create the Universe, I also took the conscious decision right at the start not to interfere in its workings. Not to respond to man's implorings, but to leave him to flounder his lonely way through. Man gets very angry with me for that. I let the innocent suffer dreadful things, I allow the guilty to go unpunished. I appear to permit a terrifyingly unjust world. Man is entitled to ask, to what purpose? Well . . . bear with me. We may come up with an answer yet.

(*The Hawking household again.* STEPHEN *wheels on in his motorised wheelchair. He wears a high collar with a tube protruding from the throat.* GOD *watches and then turns out front again.*)

Miraculous. Man is already behaving like me –
rebuilding life itself. And now we are going to
encounter yet another dimension to science's
apparently unstoppable progress. The thinking
machine. The alternative brain. This, I have to say,
could have incalculable effects on the future.

(*The* NURSE *brings on a small portable computer.
She hands it to* GOD.)

Thank you, darling.

(*She goes to* STEPHEN. GOD *looks at the audience.*)

Don't misunderstand me – she's my wife. I'm
David Mason, computer engineer and friend of the
Hawkings. My wife happens to be one of his large
coterie of nurses.

(*Goes to* STEPHEN.)

MASON   Right, Stephen, I think this will help you. I've
               managed to compress the voice synthesizer into a
               portable computer which fits onto your
               wheelchair.

               (*Attaches it to the chair and connects various
               wires. Indicates parts of the apparatus.*)

               This is the hand control. You only need two
               fingers to operate it, as with the wheelchair. The
               computer will scroll through its menu of three
               thousand key words, and you just have to pick out
               the ones you want. Like this.

               (*Demonstrates.*)

               When you've put your sentence together, just
               press like so . . . and the synthesizer does the rest.

               (*Presses a button. The synthesizer speaks in its
               characteristic disembodied American tones.*)

VOICE   Ask me a question, and I will try to answer.

MASON   Go on. You ask *me* a question.

             (STEPHEN *peers at the computer screen and plays with the control using small movements of his left fingers.* MASON *watches over his shoulder.*)

             That's it. That's the way.

             (STEPHEN *presses for the final instruction.*)

VOICE   Do I have to speak with an American accent?

MASON   (*laughing*) Sorry, Stephen. I haven't got round to giving it an English dialect yet. When I do you'll be the first to have it.

             (STEPHEN *operates the computer again, as* MASON *and the* NURSE *watch.*)

             What? What's that?

VOICE   Thank you. You have given me new life.

MASON   Well, the doctors did that – but I'm glad if I've been able to improve it a little.

             (*He turns to go. Stops. Indicates the* NURSE.)

             Is Elaine looking after you all right?

VOICE   She is a godsend.

GOD   (*to the audience, as he leaves*) Godsend. You see?

             (GOD *and the* NURSE *exit.* STEPHEN *rolls his wheelchair forwards. The backscreen shows the cosmos.*)

STEPHEN   (*his speech already prepared via the synthesizer*) As we enter the new millennium we know for certain that life on Earth is probably doomed to extinction, like the dinosaurs, within the next few million, or even thousand years. Yet our findings hint every day at other realms beyond anything we have imagined so far. Not so much the end of science, as perhaps the true beginning. Let me give you a taste of the remarkable possibilities opening before us.

(*The backscreen shows a picture of a supernova.*)

Space, we now believe, is not empty space at all. It is filled with activity.

(*The backscreen shows diagrams of superstrings.*)

Invisible matter, miniature black holes, quantum particles which can communicate from opposite ends of the Universe. Could it be therefore that the confines of space and time might be transcended in ways we never before imagined? For many physicists the idea of space-time travel is a serious scientific concept.

(*The backscreen shows a DNA molecule structure.*)

DNA. The study of genes is telling us how life itself has evolved, and may be evolving elsewhere within the myriad galaxies. And amazingly – it is bringing within our grasp the defeat of disease, of the ageing mechanism, even the very creation of life. For the first time the Darwinian process is being challenged. Trial and error no longer. This terrifies us. Our egos say, 'No! You must not mess with *our* genes' But our intelligence says, 'Make way all you imagined gods, we are taking charge at last!'

(*The backscreen shows a computer circuit.*)

Even so – physical and mental super beings we may become, yet already our lives are largely controlled by more efficient mental processes than our own. Computers can calculate the mass of a distant galaxy, can play a grand-master at chess, can guide a robot to do the housework.

(*Pointedly.*) Can even design cleverer versions of themselves.

(*Picture of a molecular memory bank.*)

We now believe it may eventually be possible to combine molecular biology with computer technology to produce the ultimate evolving brain. A self-generating intelligence that could churn its way through the mysteries of the cosmos when we are nothing but the data within the depths of its own infinite memory bank.

(*Significant pause.*)

Or is that what the cosmos already is?

(*Further pause.*)

So where does all this leave *us*? Poor uncertain, superstitious microbes that we are – what place does this give us in the unfolding wonder? That perhaps is the final question that has to be left to the philosophers. (*A beat.*) Or does it?

(*He wheels to one side. The backscreen changes to a domestic background. The* NURSE *comes on with a cup. She helps him silently to drink. She rearranges his dress, his cushions. His eyes signal.*)

NURSE   What? Tell me.

(*His fingers work at his computer control. She reads what he is saying on the screen. Smiles.*)

All right. If you behave yourself.

(*She kisses him lovingly on the forehead. Then goes.* JANE *enters from the other side. She looks after the* NURSE *for a moment. Then turns to* STEPHEN. *He composes a sentence on his computer.*)

VOICE I am so sorry.

JANE They're our friends, Stephen! How can this happen? David designed your voice for you!

VOICE Yes.

JANE The children? What about the children?

VOICE They will learn to understand.

JANE I don't see how. (*Passionate.*) *I* don't understand! After all that we've been through.

VOICE I am truly sorry.

JANE Oh, I know she gives you all the things I used to give you. She cares for you, and talks to you – answers your every need. She is your life now, but . . . does it have to mean divorce?

VOICE I don't know how much time I have. Marriage is all I can offer her.

JANE Marriage is for life! For better for worse! Jonathan and I could have taken that route but we chose not to.

VOICE That was your way.

JANE Well you may not accept the conventional codes, but surely . . .

(*She stops herself from crying.*)

Sorry, I . . . (*Pulls herself together.*) I sometimes wish you weren't so successful. Rich. I'd have to have stayed your nurse then. (*Smiles ironically.*) A brief history. Who would have thought I'd be undermined by a book.

VOICE You have your own life now.

JANE   It will always be part of your life, Stephen. We
            have gone through so much – from black holes to
            all the glittering prizes.

VOICE   Yes.

JANE   I'm sorry that . . . in the end we couldn't really
            think alike. The idea of Christ in the context of
            your great vision of everything makes little logical
            sense, I accept that . . . but to ordinary people like
            me he is something we can comprehend – relate to.
            We need that. Earthly fairy stories, as you call
            them, may seem ridiculous when you're dealing
            with distant planets, but since that is all way
            above our heads we prefer to remain here on this
            one. In one timescale. With one belief.

            (*Pause. She weeps a little.*)

VOICE   I envy you that.

            (*She pulls herself together.*)

JANE   Well – God knows what the world will make of it.

            (*Laughs drily.*)

            Sorry. God doesn't know.

            (*Goes and touches his cheek.*)

            All the same – whether you want him or not, God
            be with you.

            (*Leaves.* GOD *enters and looks after her.*)

GOD   And with you too. (*Out front, to audience.*) The
            human tragedies and comedies go on – in the
            midst of all the heaving chemistry. So what the
            hell is it all about? And I use that word
            euphemistically.

            (*Claps his hands together.*)

All right. No more games. Let me meet this challenge head on. We want some proper answers here!

(*Nods towards* STEPHEN.)

How do I confront this . . . icon of the modern world?

(*Ponders. Has an idea.*)

Remember Schrödinger's rabbit? Alive and dead in different versions of the Universe? What is to stop me – alive in my version of the Universe – challenging him to prove me dead in *his* version? Eh?

(*Turns to the backscreen and gestures. It changes to show a sweeping view of the cosmos, which could equally be a pattern of quantum particles.*)

Why shouldn't I communicate via the timeless links of his quantum world, and force him to a real showdown? Let's see what happens.

(*Turns to* STEPHEN.)

Professor Hawking.

(STEPHEN *stirs and turns his wheelchair to face* GOD.)

Still pondering the great Theory of Everything?

(STEPHEN *waits.*)

Forgive me for intruding, but it's time you and I met properly. You realise of course who I am?

(STEPHEN *starts to compose a reply on his computer.*)

There's no need for that. I think I'm allowed the occasional miracle in my universe.

(*Snaps his fingers.* STEPHEN *looks up, his body relaxes, his face loses its contortion.*)

How does that feel?

STEPHEN   Better, thank you.

GOD   (*pleased with himself*) Good. Well I just thought it was high time you, as the representative of your universe, and I, as the creator of mine, sorted out our relationship – man to god, as it were.

STEPHEN   Fine.

(GOD *paces, pondering his approach.*)

GOD   Very well – I am the God of tradition. You are Professor Stephen Hawking, C.B.E., Companion of Honour, etcetera, etcetera . . . some have said, the greatest scientific mind since Einstein. Some have said otherwise.

(STEPHEN *waits.*)

I would now like to focus on other matters than science. What is your philosophy? What do your findings tell you about the deeper things of life?

STEPHEN   That is a large question.

GOD   Then let us start with the basics. You don't believe in Christianity?

STEPHEN   I believe that Christ existed. He was a wise and gifted leader in an era that was desperate for such a one.

GOD   The right man for the right moment?

STEPHEN   Yes.

GOD   Then why do you not accept that he may have been sent by me?

STEPHEN   You hadn't interfered with the development of *homo sapiens* for a quarter of a million years. Why would you chose such an arbitrary time and such an insignificant place?

GOD   The Bible is clear. I intervened because humanity was going astray.

STEPHEN   It had always been astray. Slaughtering and stealing since it left the jungle. Did you consign all those generations to hell?

GOD   Couldn't Christ have been just one of my many interventions?

STEPHEN   Then why intervene so ineffectively?

GOD   (*put out*) Ineffective?

STEPHEN   Two thousand years after Jesus and Christianity is still struggling to find its way – along with all other religions. If you had really chosen to stick your all-powerful oar in, surely you'd have done so more emphatically?

GOD   Very well – supposing I accept that the origins of Christ himself are at any rate debatable – what of *me*?

STEPHEN   What of you?

GOD   Fundamental question. Do I, God, exist?

STEPHEN   It depends what you mean by God.

GOD   What do *you* mean?

STEPHEN   I can tell you what I don't mean.

GOD   What?

STEPHEN   A judge-like figure presiding in the clouds.

GOD   I think few people hold that image any more.

STEPHEN   Many hold with what that image represents.

GOD   Exactly! Millions of people out there pray to me
             every day.

STEPHEN   How do you persuade a woman who is screaming
             to you as her children are slaughtered in front of
             her that you answer prayer?

GOD   I may not answer it directly. I may answer it by
             giving her the strength to endure the ordeal.

STEPHEN   Why put her through it at all? You could hardly
             claim it was character forming.

GOD   So you are saying all who pray are deluded?

STEPHEN   Not at all. Prayer has a powerful subjective force.
             It is one of man's ways of taking comfort,
             focusing his moral purpose.

GOD   Ah – so science does admit a moral purpose?

STEPHEN   Of course. Some scientists still maintain a faith.
             We may have a biological existence, but we also
             have a psychological one. They both have to find
             their way for the sake of man's survival.

GOD   Then where do you suppose that moral code came
             from?

STEPHEN   It makes no sense that you ordained it. The Ten
             Commandments are a simple common-sense set of
             rules formed by society for its own self-protection.
             That's what morality is. And it evolves as society
             evolves.

GOD   So you're deducing that, if I do not interfere with
             the workings of the world, I did not create it either.

STEPHEN   Well, if you did, there are only two possibilities. Forgive me for being blunt, but either you made a botched job of it . . .

GOD   Huh!

STEPHEN   . . . or you said to yourself I will create a world where evil can fight it out with good, and see which wins. Why? For your own amusement?

GOD   As a challenge to mankind.

STEPHEN   A pretty cynical one. To what purpose – when you could have created a purely benign world to start with?

GOD   Supposing I really am up against the Devil?

STEPHEN   Ah, another mythical monster. Well, there is a demon in our heads sometimes, I grant you that.

GOD   Ah! So that is where the battle rages, in man's head?

STEPHEN   Yes. Despite his achievements he is still an extremely primitive being – just a couple of steps beyond the apes.

GOD   But surely there must be more than just this world, this life?

STEPHEN   You mean heaven, reincarnation, the afterlife?

GOD   Give me one good reason why I shouldn't have made this existence a prologue to a higher one.

STEPHEN   A dress rehearsal? To what purpose?

GOD   What purpose!

STEPHEN   We go through all the turmoil of existence and evolution here simply to be told, well these are your marks in the preliminary exam, now go to your place in the *real* Universe?

GOD   Yes! Why not?

STEPHEN   It seems a pointless exercise for an omnipotent
           deity. Why not put us all there in the first place?
           Or at least give us genuine knowledge of the
           challenge, rather than leave us floundering in a
           sea of superstition.

GOD   (*impassioned*) Come now, Professor Hawking! I
           must take personal issue with you! Man has
           accepted 'God' – in whatever form – since he first
           had constructive thought. He has created the
           greatest buildings, painted the finest pictures,
           carved the mightiest statues, formed his
           constitutions and his moral codes – all with
           reference to God. He turns to me in his deepest
           crises and in his grandest triumphs. Are you
           seriously saying he has been deluded all this
           time?

STEPHEN   Not at all.

GOD   Then what . . . ?

STEPHEN   In the absence of true knowledge about his
           environment man *has* to imagine some form of
           supreme force. He has an essential need for a
           directive, an innate yearning for a parental
           authority.

GOD   Parental?

STEPHEN   Exactly. Conditioned from the cradle. Inherited
           mythology.

GOD   (*outraged*) What??

STEPHEN   But he is growing up. For the first era in his
           history, man is now starting to comprehend the
           true workings of the Universe – which are more
           awe-inspiring than anything he has conceived so
           far. With that knowledge he no longer has to
           fantasise about his destiny.

GOD	He knows his destiny?
STEPHEN	Beginning to guess at it.
GOD	What is it?
STEPHEN	To fulfil that Universe's promise. To work at its progress.
GOD	The atom bomb. Is that progress?
STEPHEN	Nuclear fission can destroy cities or power them. It is up to man.
GOD	So ultimately science has all the answers.
STEPHEN	Let's just say, it will open the gates to all the questions. When you see what science has achieved in a hundred years, imagine what it will achieve in the next ten thousand.
GOD	If man survives that long.
STEPHEN	The more science progresses, the better his chances.
GOD	And what about love? Is there a place for old-fashioned love in this great march of science?
STEPHEN	Fundamentally. It is the supreme expression of unity. And unity is the ultimate destination.
GOD	It's a grand and sweeping philosophy, Professor! But tell me – without God, what place does the poor humble individual have in all this mighty progress?
STEPHEN	Every act, every gesture he makes either advances or impedes the inexorable march. The future is composed of the individual's contributions.
GOD	But *man* is not God?

STEPHEN   Of course not. He is essentially fallible still.

GOD   (*triumphant*) And you are a man!

STEPHEN   (*after a moment*) Yes. I am fallible. I have made
            many mistakes in my life.

GOD   So – if the fallible you says I did not imagine the
            Universe in the first place, who did?

STEPHEN   If you – who imagined *you* in the first place?

GOD   Ah.

STEPHEN   You see? It's no solution to answer one mystery
            by simply creating another.

GOD   So that's it? That's your final conclusion? The
            Mind of God – as you yourself have called it, is
            simply a figment of the *human* mind?

STEPHEN   Ah. I have not said that.

GOD   Then what have you said?

STEPHEN   As I've implied, the word 'God' can be whatever
            we interpret it to mean. The final consequence.
            The initial impulse. The ultimate fulfilment. Until
            we come to it, it has to remain in our imaginations.

GOD   And the God in *your* imagination?

            (*Pause.* STEPHEN *has no immediate response.*)

STEPHEN   I . . . I have not defined it yet.

GOD   Ah.

STEPHEN   But I know that it is there, inherent in the endless
            experiment of the Universe. A solution that –
            unlike all the metaphysical theories and creeds –
            will eventually seem so obvious . . . so self-
            evident . . . we will realise it was part of us all
            along.

God   Part of you . . .

Stephen   Us. Our Universe. Not yours.

               (Stephen *turns his chair and wheels away. Theme music. The lights dim.* God *comes slowly towards the audience.*)

God   I'm not sure how I came out of that. I'm not sure whether we've resolved anything at all. Like Schrödinger's rabbit, I still feel both dead and alive at once.

               (*A beat.*)

               Well, that's pretty pathetic, I can hear you thinking. Is that all you have to offer us – a half-dead rabbit? And not even pulled out of a hat! We thought we were going to get some answers here, you're thinking. We thought you were going to come up with *something* persuasive after making us sit through all that!

               (*Puts up a finger. Smiles.*)

               Of course I am. You didn't think I'd just leave it at that, did you? You should know me better by now.

               (*Looks off after* Stephen.)

               I'm going to have the last word, whatever he thinks.

               (*Looks back.*)

               Let me offer you a supposition – beyond perhaps your farthest imaginings. Suppose that he and all his fellow geniuses are right. Suppose that your universe is indeed self-creating, and unfolds its secrets over an infinite cycle of years and light-years. Suppose that man, or machine, or some evolved super-intelligence eventually makes it to

the ultimate non-boundaries of space and time.
What then?

(*The picture of the cosmos swirls and expands.*)

As we have seen, that universe is unquantifiable.
Light – space – time – matter – they are all merely
relative concepts. All just expressions of one
basic concept – energy. And what is energy?
Simply the compulsion to *be*.

(*Pause. He speaks slowly and deliberately.*)

Suppose therefore that the life of the Universe *is* a
complete and total entity. Originating from
nothing – striving to create something –
struggling through trial and error to first gain
movement, then matter, then life, then thought,
then knowledge . . . until finally it attains the
supreme objective of progress . . .

(*The cosmos merges into a picture of a gigantic
spiral galaxy.*)

It completes the great circle, and evolves back to
nothing again. Except that by now that nothing is
its ultimate achievement of *everything*. Total
knowledge – of itself. Total joy – in itself. Total
love – of itself.

(*Pause. The spiral galaxy rotates and dissolves to
an infinite, dazzling whiteness.*)

In other words . . . perfection.

(*Pause.*)

In other words . . . me.

(*Smiles and goes.*)

CURTAIN